FRONTIERS IN *Staphylococcus aureus*

Edited by **Shymaa Enany**
and **Laura E. Crotty Alexander**

Frontiers in *Staphylococcus aureus*
http://dx.doi.org/10.5772/63039
Edited by Shymaa Enany and Laura E. Crotty Alexander

Contributors

Irma Isnafia Arief, Kotaro Suzuki, Kidon Sung, Jungwhan Chon, Saeed Khan, Szweda Piotr, Barbara Kot, Jaime López-Cervantes, Andres Chavez Almanza, Ernesto Uriel Cantú-Soto, Chi Zhang, Xiaoxiao Wu, Dima Youssef, Kate Molony, Shymaa Enany, Laura Elise Crotty Alexander, Elisa McEachern, Juan José Valdez-Alarcón, Claudia Ibeth Barrera-Rivas, Norma Anahí Valle-Hurtado, Graciela M. González-Lugo, Víctor Manuel Baizabal-Aguirre, Alejandro Bravo-Patiño, Marcos Cajero-Juárez, Arumugam Gnanamani, Periasamy Hariharan, Maneesh Paul-Satyaseela

Published by InTech
Janeza Trdine 9, 51000 Rijeka, Croatia

Publishing Process Manager Martina Usljebrka
Technical Editor SPi Global
Cover InTech Design team

First published March, 2017
Printed in Croatia
Legal deposit, Croatia: National and University Library in Zagreb

Additional hard copies can be obtained from orders@intechopen.com

Frontiers in *Staphylococcus aureus,* Edited by Shymaa Enany and Laura E. Crotty Alexander
p. cm.
Print ISBN 978-953-51-2981-3
Online ISBN 978-953-51-2982-0

For EU product safety concerns:
IN TECH d.o.o., Prolaz Marije Krucifikse Kozulić 3, 51000 Rijeka, Croatia,
info@intechopen.com or visit our website at intechopen.com.

Meet the editors

Dr. Shymaa Enany is an assistant professor of Microbiology and Immunology at the Suez Canal University, Egypt. She received her PhD degree from the School of Medical and Dental Sciences, Niigata University, Japan. Her dissertation focused on the molecular characterization of community-acquired methicillin-resistant *Staphylococcus aureus* (CA-MRSA), the leading cause of many life-threatening illnesses. For her postdoctoral work, she collaborated with many laboratories in San Diego, California, USA, and in Niigata, Japan, analyzing the proteome of *Staphylococcus aureus* and monitoring the reactions of innate immunity and airway cytokines in response to MRSA infections.

Dr. Laura E. Crotty Alexander is a Pulmonary Critical Care physician at the University of California, San Diego, and VA San Diego Healthcare System. As a biomedical basic science researcher, she studies host-pathogen interactions at molecular, cellular, organ, and organism levels and recently defined pro-virulent effects of inhalants such as cigarette smoke and e-cigarette vapor on Staphylococcal virulence. Dr. Crotty Alexander's clinical interests include bacterial pneumonia and sepsis, e-cigarette effects on innate immunity, and moderate-to-severe asthma.

Contents

Preface

Staphylococcus aureus is a beautiful golden bacterium that haunts epithelial surfaces of the human body, patiently awaiting the opportunity to invade and infect even the healthiest individuals. What makes it successful as a human pathogen? How has it developed antibiotic resistance so rapidly? And, how has it been able to spread across the world so swiftly? Here, multiple experts examine this golden toxin-producing Gram-positive organism and provide insight into what genotypic and phenotypic shifts have occurred in *S. aureus* over the past 100 years to make it the superbug it is today.

There were several underlying reasons why we wrote this book. Firstly, rates of significant *S. aureus* clinical infections have continued to climb. Secondly, defining various modes of *S. aureus* transmission may lead to improved prevention and increased recognition of *S. aureus* infections. Thirdly, the need for discovering new ways to rapidly detect this aggressive organism has become an urgent issue.

Through the chapters within, the authors examine patterns of colonization and exposures in humans, mammals, and birds that have led to the development of antibiotic resistance, including methicillin-resistant *S. aureus* (MRSA). Alternative, novel chemotherapeutics to target *S. aureus* are discussed, including herbal medicines, bee products, and modes of delivery, as conventional antibiotic options to treat this aggressive, multifaceted, and readily adaptable pathogen are becoming limited. This book is an excellent starting point for anyone who wants or needs to study *S. aureus*. Most of the chapters are oriented toward the detection and prevention of *S. aureus* and research for unconventional treatment of patients infected with these bacteria, and importantly, we include several review chapters to allow scientists and clinicians to better understand the epidemiology, transmission, and clinical significance of these golden bacteria.

We believe that our "Frontiers in *Staphylococcus aureus*" is an excellent book for microbiologists, especially those who are interested in this superbug. We hope you enjoy reading it. Finally, we would like to thank all the contributing authors who contributed a great deal of time and original research to this project.

Dr. Shymaa Enany
Department of Microbiology and Immunology, Faculty of Pharmacy
Suez Canal University, Ismailia, Egypt

Dr. Laura E. Crotty Alexander
Department of Medicine, University of California, San Diego,
and VA San Diego Healthcare System, San Diego, CA, USA

Section 1

Staphylococcus aureus Overview

Chapter 1

Staphylococcus aureus: Overview of Bacteriology, Clinical Diseases, Epidemiology, Antibiotic Resistance and Therapeutic Approach

Arumugam Gnanamani, Periasamy Hariharan and Maneesh Paul-Satyaseela

Additional information is available at the end of the chapter

http://dx.doi.org/10.5772/67338

Abstract

Staphylococcus aureus is an important human pathogen that causes wide range of infectious conditions both in nosocomial and community settings. The Gram-positive pathogen is armed with battery of virulence factors that facilitate to establish infections in the hosts. The organism is well known for its ability to acquire resistance to various antibiotic classes. The emergence and spread of methicillin-resistant *S. aureus* (MRSA) strains which are often multi-drug resistant in hospitals and subsequently in community resulted in significant mortality and morbidity. The epidemiology of MRSA has been evolving since its initial outbreak which necessitates a comprehensive medical approach to tackle this pathogen. Vancomycin has been the drug of choice for years but its utility was challenged by the emergence of resistance. In the last 10 years or so, newer anti-MRSA antibiotics were approved for clinical use. However, being notorious for developing antibiotic resistance, there is a continuous need for exploring novel anti-MRSA agents from various sources including plants and evaluation of non-antibiotic approaches.

Keywords: *Staphylococcus aureus*, MRSA, CA-MRSA, HA-MRSA, anti-MRSA

1. Introduction

Staphylococcus aureus is a Gram-positive bacterium and causative agent of wide range of infectious diseases such as skin infections, bacteremia, endocarditis, pneumonia and food poisoning. The organism was originally a leading nosocomial pathogen and afterwards epidemiologically distinct clones emerged in community settings. *S. aureus* expresses number of virulence factors which help to establish infection by facilitating tissue attachment, tissue invasion and evading from host immune response. The ability to acquire resistance to multiple antibiotics classes makes *S. aureus*, a challenging pathogen to treat. Emergence and spread of *S. aureus* strains which are resistant to methicillin, referred to as methicillin-resistant *S. aureus* (MRSA) resulted in high morbidity, high mortality and increased treatment costs. Vancomycin remained gold standard drug to tackle these strains for years but the emergence of resistance restricted its clinical utility. Newer anti-MRSA antibiotics which were approved by U.S. FDA came as respite for clinicians. However, new antibiotic discovery efforts and non- antibiotic approaches to tackle MRSA should not be diminished considering the ability of the pathogen to acquire resistance to newer drugs quickly after their introduction in clinics.

In this chapter, we present a comprehensive outlook of *S. aureus* with account on bacteriology, pathogenesis, epidemiology, antibiotic resistance and therapeutic approaches.

2. Bacteriology

2.1. Microscopic morphology

S. aureus cells are Gram-positive and appear in spherical shape. They are often in clusters resembling bunch of grapes when observed under light microscope after Gram staining. The name 'Staphylococcus' was derived from Greek, meaning bunch of grapes (*staphyle*) and berry (*kokkos*) [1]. The scanning electron microscopic observation reveals roughly spherical shaped cells with smooth surface [2]. The diameter of the cells ranges from 0.5 to 1.0 μM [3]. The transmission electron microscopy of cells shows thick cells wall, distinctive cytoplasmic membrane and amorphous cytoplasm [4].

2.2. General cultural and biochemical characteristics

S. aureus is an aerobic and facultative anaerobic organism that forms fairly large yellow or white colonies on nutrient rich agar media. The yellow colour of the colonies is imparted by carotenoids produced by the organism. The term 'aureus' is derived from Latin, which refers to the colour of gold [5]. The organism is often haemolytic in blood agar due to production of four types of haemolysins (alpha, beta, gamma and delta) [6, 7]. Nearly all isolates of *S. aureus* produce coagulase enzyme, a virulence factor that also helps in identification of the organism [6, 8]. The organism is salt tolerant, which is able to grow in mannitol-salt agar medium containing 7.5% sodium chloride [8]. The organism is catalase positive and oxidase negative.

2.3. Medical laboratory diagnosis

The primary objective in laboratory diagnosis is to identify whether the diagnosed *S. aureus* isolate is methicillin resistant. Since MRSA emerged as problematic pathogen, a systematic diagnostic approach is necessary for early diagnosis so that treatment with appropriate antibiotics can be initiated as early as possible. For the species identification, slide and tube

http://dx.doi.org/10.5772/67338

coagulase tests, latex agglutination tests and PCR-based tests are used. For detection of MRSA, determination of minimum inhibitory concentration (MIC) of methicillin or oxacillin or cefoxitin using broth micro-dilution method, cefoxitin disk screen, oxacillin agar screen and latex agglutination test for PBP2a and molecular methods for detection of *mecA* are employed [8].

3. General pathogenesis and clinical diseases

3.1. Pathogenesis

The process of *S. aureus* infections involves five stages. They are (1) colonization, (2) local infection, (3) systemic dissemination and/or sepsis, (4) metastatic infections and (5) toxinosis. The organism is in carrier state in the anterior nares and can remain so without causing infections for weeks or months. The colonization proceeds to infection under certain predisposing factors such as prolonged hospitalization, immune suppression, surgeries, use of invasive medical devices and chronic metabolic diseases. Localized skin abscess develop when the organism is inoculated into the skin from a site of carriage. This can further spread and results in various clinical manifestations of localized infections such as carbuncle, cellulitis, impetigo bullosa or wound infection. The organism can enter into blood and spread systemically to different organs causing sepsis. This haematogenous spread may result in endocarditis, osteomyelitis, renal carbuncle, septic arthritis and epidural abscess. Without a blood stream infection, specific syndromes can occur due to extra cellular toxins of *S. aureus*. These are toxic shock syndrome, scalded skin syndrome and foot borne gastroenteritis [9].

3.2. Hospital and community infections

S. aureus causes wide range of infections in human. The clinical infections of *S. aureus* are classified into community and nosocomial categories based on origin of infection. These two types are distinct in clinical manifestations of the infections, antibiotic susceptibility and the genetic background of the infecting *S. aureus* strains. For decades, *S. aureus* has been predominately a nosocomial pathogen and is a leading cause of mortality and morbidity in hospitals. However, the community *S. aureus* infections are in rise. The important clinical *S. aureus* infections are bacteraemia, infective endocarditis, skin and soft tissue infections, osteoarticular infections and pleuropulmonary infections. Other clinical infections are epidural abscess, meningitis, toxic shock syndrome and urinary tract infections [9, 10].

3.3. Virulence factors

S. aureus possess battery of virulence factors. These factors enable the organism to be successful as pathogen that causes wide range of human and animal infections. Virulence factors help in attachment to host cells, breaking down the host immune shield, tissue invasion, causing sepsis and elicit toxin-mediated syndromes. This is the basis for persistent staphylococcal infections without strong host immune response [11]. Based on their mechanism of action and role in pathogenesis, staphylococcal virulence factors are classified as represented in **Table 1** [9, 12].

Factors	Characteristics
Helping attachment to host tissues	
Microbial Surface Components Recognizing adhesive matrix molecules (MSCRAMM)	Cell surface proteins which interact with host molecules such as collagen, fibronectin & fibrinogen, thus, facilitate the tissue attachment. Staphylococcal protein A, fibronectin-binding proteins A and B, collagen-binding protein & clumping factor A & B belong to this family. They are also involved in host immune evasion [13].
Breaking/evading the host immunity	
Polysaccharide microcapsule	Resist the phagocytosis & killing by polymorphonuclear phagocyte [14].
Protein A	It binds to Fc portion of immunoglobulin, prevents opsonization, functions as super antigen & limits the host immune response [15].
Panton-Valentine leukocidin (PVL)	PVL is found in most of community-associated MRSA (CA-MRSA) [16]. PVL belongs to group of membrane pores forming proteins. It consists of two protein components (LukS-PV and LukF-PV) which act together as subunits and form porins on cell membrane of host cells, leading to leakage of cell contents and cell death [17].
Alpha-toxin (Alpha hemolysin)	It was the first bacterial exotoxin to be identified as a cell membrane pore former which causes cell leakage & death [18].
Chemotaxis-inhibitory protein of S. aureus (CHIPS):	CHIPS is an extracellular protein which inhibits the chemotaxis functioning of neutrophil and monocytes [19].
Tissue invasion	
Extracellular adherence protein (Eap)	An exoprotein which binds to host cell matrix, plasma proteins & endothelial cell adhesion molecule ICAM-1. In addition to the roles of adhesion and invasion, it also has immune-modulatory activity [20].
Proteases, lipases, nucleases, hyaluronatelyase, phospholipase C, metalloproteases (elastase), & Staphylokinase	These extracellular enzymes cause tissue destruction and, thereby, help in bacterial penetration into tissues.
Induces toxinosis	
Enterotoxins	*S. aureus* produces battery of enterotoxins which are potent gastrointestinal exotoxins. The Staphylococcal food poisoning is an intoxication which results from consumption of foods containing sufficient amount of preformed enterotoxins [21].
Toxic shock syndrome toxin -1 (TSST-1)	TSST-1 & some of enterotoxins are called as pyrogenic toxin super antigens. TSST-1 causes toxic shock syndrome especially in menstrual women [7].
Exfoliative toxins A and B	Serine proteases which selectively recognize and hydrolyze desmosomal proteins in the skin. ETs cause staphylococca-scalded skin syndrome, a disease predominantly affecting infants [22].

Table 1. Virulence factors of *S. aureus* and its characteristics.

4. Epidemiology of infections

4.1. Nasal carriage

S. aureus is a commensal and opportunistic pathogen. The anterior nares are the principal ecological niche, where the organism colonizes in humans. The nasal carriage of *S. aureus* increases the risk of infection especially in the hospital settings [23]. The average nasal carriage of *S. aureus* could be at 30% of human population [24]. Since, the nasal carriage increases the risk of development of surgical site, lower respiratory and blood stream infections in hospitals, efforts are made to eliminate the carriage using various strategies. Methods such as local application of antibiotics (eg. mupirocin) or disinfectants, administration of systemic antibiotics and use of a harmless *S. aureus* strain (type 502A) which competes for the colonization of nares with existing one are employed to decolonize the *S. aureus* from nares [25–28].

4.2. Emergence and evolution of MRSA

The MRSA are those *S. aureus* strains carrying a *mecA* gene, which codes for additional penicillin-binding protein, PBP2a. The beta-lactam antibiotics exert their antibacterial activity by inactivation of penicillin-binding proteins (PBPs), which are essential enzymes for bacterial cell wall synthesis. However, these antibiotics have only a low affinity towards PBP2a, thus this enzyme evades from inactivation and carry out the role of essential PBPs resulting in cell wall synthesis and survival of bacteria even in presence of beta-lactam antibiotics. Due to the presence of *mecA*, MRSA are resistant to nearly all beta-lactam antibiotics [29].

Penicillin is the first beta-lactam antibiotic discovered in 1928 and found to be effective weapon against *S. aureus* infections. In 1940s, sooner after its introduction into clinics, there were reports of *S. aureus* strains that were resistant to penicillin [30]. These strains produced plasmid-encoded beta-lactamase enzyme (penicillinase) which enzymatically cleaved the beta-lactam ring of penicillin rendering the antibiotic inactive [31, 32]. In 1950s, the penicillin resistance was restricted to hospital isolates of *S. aureus*. By late 1960s, more than 80% *S. aureus* isolates, irrespective of community and hospital origin, were resistant to penicillin due to plasmid transfer of penicillinase gene (*blaZ*) and clonal dissemination of resistant strains [33, 34].

Meanwhile, scientists who were challenged with penicillinase-mediated resistance in *S. aureus* discovered methicillin, a semi-synthetic penicillin that withstood the enzymatic degradation of penicillinase. Methicillin was introduced into clinics in 1961; however, in less than a year, resistance of *S. aureus* isolates to methicillin (MRSA) was reported [35]. Over the next 10 years, increasing number of MRSA outbreaks was reported in different parts of the world especially from the European countries [36, 37]. The notable feature of these reports is that, the incidences were from hospitals and thus MRSA emerged as a hospital-borne pathogen. The mechanism of resistance to beta-lactam antibiotics in these MRSA isolates was uncovered in 1981 [38].

As mentioned earlier, MRSA isolates carry a gene *mec A* which codes for PBP2a. The gene is part of a 21–60 kb mobile genetic element referred to as staphylococcal cassette chromosome *mecA* (SCC*mecA*). There are two hypotheses that explain the evolutionary origin of MRSA. The single clone hypothesis suggests that the mobile genetic element entered the *S. aureus* population on one occasion and resulted in the formation of a single MRSA clone that has since spread around the world. The second and the most agreed hypothesis is that MRSA strains evolved number of times by means of the horizontal transfer of the mobile genetic element into phylogenetically distinct methicillin-susceptible *S. aureus* (MSSA) precursor strains [39, 40].

SCC*mec* elements are highly diverse in their structural organization and genetic content (**Figure 1**) and have been classified into types based on the combination of *mec* and *ccr*, which share variations (five classes in *mec* and eight in *ccr*). To date, at least 11 types of *SCCmec* elements have been identified [41–43].

4.3. Health care-associated and community MRSA

4.3.1. Health care-associated MRSA (HA-MRSA)

Health care-associated MRSA (HA-MRSA) are those *S. aureus* isolates obtained from patients 2 or more days after hospitalization or with the MRSA risk factors (history of recent hospitalization, surgery, dialysis, or residence in a long-term care facility within 1 year before the MRSA-culture date or presence of a permanent indwelling catheter or percutaneous medical device (e.g. tracheostomy tube, gastrostomy tube or Foley catheter) at the time of culture or previous isolation of MRSA [44, 45]. Community-associated MRSA (CA-MRSA) are those *S. aureus* isolates obtained from patients within 2 days of hospitalization and without the above-mentioned MRSA risk factors.

Till 1990s, MRSA isolates were predominantly HA-MRSA and were also resistant to non-beta-lactam antibiotics. The multi-drug resistant phenotype of HA-MRSA was due to presence of non-beta-lactam antibiotic-resistant determinants in relatively large SCC*mec* [46]. During the period of 1960s to early 1990s, number of clones of HA-MRSA had spread widely across the world and HA-MRSA became endemic in hospitals and emerged as leading nosocomial pathogen [47]. The genetic background of these MRSA clones was characterized initially using phage typing subsequently by multilocus sequence typing (MLST), pulsed-field gel electrophoresis (PFGE), *spa* typing and SCC*mec* typing. The analysis of the genetic background of HR-MRSA

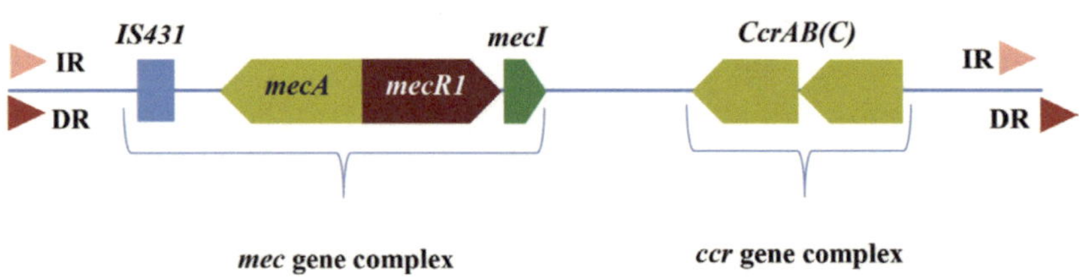

Figure 1. Basic structure of SCC*mec*. SCC*mec* constituted by *mec* gene and *ccr* gene complexes. The *mec* gene complex encodes PBP2a (*mecA*) and resistance regulators (*mecI* and *mcR1*). The *ccr* gene complex encodes the integration and excision of entire SCC element. The gene complexes are flanked by characteristic nucleotide sequences, inverted repeats (IR) and direct repeats (DR), at both ends. J (joining) regions are J1 (between right chromosomal junction and *ccr* complex, J2 (between *ccr* and *mec* complexes) and J3 (between *mec* complex and left chromosomal junctions). Adopted from Ref. [41].

isolates using these methods revealed the spread of early MRSA clone (Archaic clone) which contained type I SCC*mec* and sequence type 250 (ST250) in 1960s and extended into the 1970s in the form of Iberian clone. The Iberian clone was sequence type 247 (ST247) which evolved from ST250-MRSA by a single point mutation [48]. In the mid to late 1970s, Archaic and Iberian MRSA clones declined while, clones with novel SCC*mec* types II and III had emerged marking the on-going worldwide pandemic of HA-MRSA in hospitals and health care facilities [49, 50]. The lineages of common HA-MRSA clones are represented in **Table 2**. The rise in the prevalence of HA-MRSA throughout the world has been dramatic. In the United States, the proportion of MRSA among *S. aureus* isolates from the hospitalized patients was 2.4% in 1975, which increased to 51.6% (ICU patients) and 42% (non-ICU inpatients) by 1998–2003. Similar persistently high or increasing rates of MRSA among *S. aureus* isolates have also been observed for health care settings in many other regions of the world [51].

4.3.2. *Community-associated MRSA (CA-MRSA)*

MRSA isolates obtained from outpatients or from patients within 48 h of hospitalization and if they lack HA-MRSA risk factors mentioned earlier are referred to as CA-MRSA [52]. Scattered case reports of MRSA infections in healthy population whom had no exposure to health care facilities were published in the 1980s and mid-1990s. Beginning in 1993, case series of MRSA infection and colonization of patients lacking health care-associated risk factors were reported from six continents, in diverse states, nations and regions [51, 53]. The phenotypic and genotypic characterization of CA-MRSA isolates revealed the differences between CA-MRSA and HA-MRSA strains. While HA-MRSA strains carried a relatively large SCC*mec*, belonging to type I, II or III, CA-MRSA strains carried smaller SCC*mec* elements, most commonly type IV or type V. HA-MRSA strains were resistant to many classes of non-beta-lactam antibiotics, thus display multi-drug resistant phenotypes. CA-MRSA strains were often sensitive to non-beta-lactam antibiotics. Another notable feature of CA-MRSA strains was presence of genes for the PVL, which was rare among the HA-MRSAs. With respect to clinical cases, CA-MRSA infections were prevalent in previously healthy younger patients in contrast to HA-MRSA, which cause infections in hospitalized patients. CA-MRSA was often associated with skin and skin structure infections while HA-MRSA was implicated in wide range of infections such as pneumonia, bacteraemia, and invasive infections [48, 51]. Compared to infections caused by HA-MRSA, CA-MRSA infections had been associated with fulminant and lethal infections and worse clinical outcomes [49, 53].

Among the various clones of CA-MRSA, ST93, ST80 and ST8 are presently the predominant clones in Australia, Europe and the United States, respectively. In the United States, ST8-USA 300 is the most wide spread CA-MRSA clone [54], which harbour SCC*mec* type IV and genes encoding PVL. The concern about this clone is high virulence and increase in resistance to non-beta-lactam antibiotics [50, 53]. In United Kingdom, EMRSA-15 (ST22) and EMRSA-16 (ST36) are the dominant clones [49]. In Europe, ST80-IV, ST8-IV, ST398-V and ST152-V were commonly reported [55]. In Mediterranean countries, the dominant clones are ST80-IV and ST5-IV/V [55, 56].

In the last 10 years, there is a dramatic change in epidemiology of CA-MRSA as they invaded the health care settings. In 2008, first case of MRSA isolated from hospitalized patient turned out to

Clonal complex	Molecular sequence type	Common names for specific MRSA clones	Comment
CC5	ST5	USA100 and NewYork/Japan clone	Most common US health care-associated MRSA, SCCmecII
	ST5	EMRSA-3	SCCmecI
	ST5	USA800/Pediatric clone	Prevalent in Argentina, Colombia, United States, SCCmecIV
	ST5	HDE288/Pediatric clone	SCCmecVI
CC8	ST250	Archiac	First MRSA clone identified, COL strain as an example; SCCmecI
	ST247	Iberian clone and EMRSA-5	Descendant of COL-type strains, SCCmecIII
	ST239	Brazilian/Hungarian clone	SCCmecIII
	ST239	EMRSA-1	Eastern Australian epidemic clone of 1980s, SCCmecIII
	ST8	AUS-2 and Aus-3	SCCmecII
	ST8	Irish-1	Common nosocomial isolate in the 1990s in Europe and the United States
	ST8	USA500 and EMRSA-2-6	SCCmecIV
CC22	ST22	EMRSA-15	International clone, prominent in Europe and Australia, SCCmecIV
CC30	ST36	USA200 and EMRSA-16	Single most abundant cause of MRSA infections in UK; second most common cause of MRSA infections in US hopsitals in 2003, SCCmecII
CC45	ST45	USA600 and Berlin	SCCmecII

Table 2. The lineages of common HA-MRSA (based on Ref. [49]).

be a CA-MRSA which marked the arrival of CA-MRSA into nosocomial settings [57]. Since then, hospital outbreaks of *S. aureus* strains which are phenotypically and genotypically CA-MRSA, have been reported many parts of the world [55]. Entry of CA-MRSA into hospitals blurred the differences between CA-MRSA and HA-MRSA. The increased reports of CA-MRSA outbreaks in hospital suggest that CA-MRSA may eventually displace HA-MRSA in hospitals [58].

http://dx.doi.org/10.5772/67338

5. Antibiotic resistance

5.1. Beta-lactam resistance

5.1.1. Penicillin resistance

The first beta-lactam antibiotic penicillin G was discovered in 1928 by Alexander Fleming and the drug was used in human as chemotherapeutic agent in 1941 [59]. The antibiotic was potent against Gram positive pathogens [60] and a power weapon against Staphylococcal infections. However, first reports of *S. aureus* strains that were resistant to penicillin appeared after a year of its clinical use [30]. Such penicillin-resistant isolates carried a plasmid gene, *blaZ* which encoded a beta-lactamase enzyme, referred to as penicillinase [33, 34]. The enzyme is capable of cleaving the beta-lactam ring of penicillin resulting inactivation of the antibiotic [31, 32].

The emergence and spread of penicillinase-mediated resistance in *S. aureus* is referred to as first wave of resistance. This has spread in alarm proportions and became pandemic in the 1960s. About 80% of both community and hospital acquired *S. aureus* isolates were resistant to penicillin by late 1960s [33, 49]. By early 2000s, more than 90% of Staphylococcal isolates produced penicillinase enzyme irrespective of their community or hospital origin [34].

5.1.2. Methicillin resistance

As discussed earlier, the penicillinase resistance in *S. aureus* was countered by the discovery of methicillin, penicillinase-stable semisynthetic penicillin. The drug was introduced into clinics in 1961 and subsequently strains showing methicillin resistance (MRSA) was reported in the same year [35]. After the initial report, MRSA clones spread rapidly across the world but restricted to nosocomial settings. This is referred to as second wave of beta-lactam resistance in *S. aureus* [40]. As discussed earlier, methicillin resistance was mediated by the presence of *mecA* gene. The therapeutic outcome of MRSA infections was worse than methicillin sensitive *S. aureus* (MSSA) due to the underlying comorbid factors such as old age, immune suppression and, importantly, lack of effective antibiotics to treat MRSA, which were often multidrug resistant [34]. The rise in MRSA infections in hospitals resulted in high morbidity and mortality and increase in cost of health care [61, 62].

The third wave of beta-lactam resistance in *S. aureus* began with reports of MRSA infections in community in early 1990s. As discussed earlier, these strains were phenotypically and genetically distinct from MRSA isolates from hospitalized patients, resulting in definitions of HA-MRSA and CA-MRSA [51, 53]. In the last decade, community MRSA strains invaded the hospital settings and the difference between HA and CA MRSA is now blurred [58].

5.2. Quinolones resistance

Nalidixic acid, the prototype quinolone and the second generation quinolones (e.g. ciprofloxacin and norfloxacin) are predominately active towards Gram negative bacteria while third generation (e.g. levofloxacin) and fourth generation (e.g. moxifloxacin, gemifloxacin) quinolones exhibited improved and greater activity against Gram-positive bacteria [63–65]. Quinolones exert their antibacterial action by inhibiting bacterial topoisomerases (topoisomerase IV and DNA Gyrase), which are essential for relieving DNA super coiling and separation of concatenated DNA strands [66]. The resistance to quinolones in *S. aureus* arises in stepwise manner, due to point mutations primarily in GrlA subunit of topoisomerase IV and GyrA subunit of Gyrase. Additional mechanism by which *S. aureus* become resistant to quinolones is by expression of NorA efflux pumps [67].

The quinolone resistance in *S. aureus* is mostly associated with methicillin resistance though the mechanism of resistance and encoding genes are altogether different from each other. This could be due to higher usage of quinolones in hospital settings where the HA-MRSA prevalence is high resulting in selection of quinolone resistance [68–70]. In year 2008, the fluoroquinolone resistance among MRSA isolates implicated in acute bacterial skin and skin structure infections (ABSSSIs) in hospitals was at 70.3%. Due to such high level of quinolone resistance among MRSA in hospital settings, even third- and fourth-generation quinolones have not been considered for treatment of MRSA [71]. With respect to CA-MRSA, though they were susceptible to non-beta-lactam antibiotics including quinolones, the scenario has changed in recent years, with the rise in incidence of CA-MRSA infections which were multi-drug resistant [72].

5.3. Vancomycin resistance

Vancomycin, a glycopeptide antibiotic, was discovered from a microbial source (*Streptomyces orientalis*) in 1952. The drug was approved for clinical use in 1958; however, it was eclipsed by methicillin and other anti-staphylococcal penicillins which were considered less toxic than vancomycin and equally efficacious against penicillin-resistant Staphylococci [73]. Beginning early 1980s, there was sudden increase in vancomycin usage due to rise in HA-MRSA infections and emergence of pseudomembranous enterocolitis cause by *Clostridium difficile* in hospitalized patients [73–75]. Clinical efficacy of vancomycin efficacy in treatment of MRSA infections was well established over the period of time, thus the drug emerged as workhorse anti-MRSA drug [76].

5.3.1. Vancomycin intermediate S. aureus

The antibacterial activity of vancomycin is mediated by its binding to the C-terminal D-Ala-D-Ala residue of the peptidoglycan precursor, and formation of non-covalent complex, thereby, prevents the use of the precursor in bacterial cell wall synthesis [77, 78]. Three decades after its introduction into clinics, no clinical resistance to vancomycin was reported. The first report of a MRSA strain showing reduced susceptibility to vancomycin was reported in 1997. The vancomycin MIC against this strain (Mu50) was 8 mg/L, thus, designated as intermediate sensitive category. The strain had thickened cell wall when observed under electron microscopy and did not carry *vanA* or *vanB* genes as found in vancomycin-resistant enterococci (VRE) [79]. Subsequently, there were more reports of clinical infections due to MRSA strains with decreased vancomycin susceptibility similar to that of Mu50 strain. The *S. aureus* strains with a MIC range of 4–8 mg/L are referred to as vancomycin intermediate *S. aureus* (VISA). There were strains, which showed vancomycin MIC of 2 mg/L but had subpopulation with vancomycin MIC of 4–8 mg/L. These strains are referred to as hetero VISA (hVISA) [80, 81].

The genetic basis of emergence of VISA appears complex. The genetic analysis of VISA strains identified mutations in determinants that control the biosynthesis of bacterial cell wall and/or mutations in the ribosomal gene rpoB [82]. The increased MRSA infection in hospitals has led to extensive use of vancomycin resulting in the selection of MRSA strains with reduced vancomycin susceptibility [83]. The study on prevalence of hVISA and VISA has met with the problem of accurate detection of decreased susceptibility to vancomycin. Different diagnostic methods showed variable sensitivity and specificity leading to contradictory reports in prevalence [80, 84–86]. During 2010–2014, the prevalence rates of hVISA and VISA among MRSA strain were at 7.01% and 7.93%, respectively [87]. The emergence and increased incidence of hVISA and VISA has limited the therapeutic use of vancomycin in the treatment of MRSA infections in hospital. However, by optimizing the dose regimen and drug delivery, thereby, achieving the desired blood plasma concentration which would give the clinical efficacy is the way forward in preserving the clinical utility of vancomycin [88, 89].

5.3.2. *Vancomycin-resistant S. aureus*

S. aureus strains which are referred to as hVISA and VISA are not considered resistant based on vancomycin susceptibility breakpoint (vancomycin MIC of 8 mg/L) defined by clinical laboratory standards institute (CLSI). Unlike VRE, these strains do not carry *vanA* or *vanB* type of genes to confer resistance to vancomycin. In 2002, first report of a *S. aureus* strain showing vancomycin MIC of >128 mg/L was published. The strain was methicillin resistant and carried *vanA* gene which was responsible for high-level resistance to vancomycin [90]. This report was followed by sporadic incidences of isolation of *S. aureus* strains with resistance to vancomycin [91]. All these strains showed high vancomycin MIC (>8 mg/L) and are referred to as vancomycin-resistant *S. aureus* (VRSA).

VRSA strains carried copies of the transposon Tn*1546*, which was acquired from vancomycin-resistant *Enterococcus faecalis*. The transposon which mediates the VanA-type resistance, encodes a dehydrogenase (VanH), which reduces pyruvate to D-Lac, and the VanA ligase, which catalyzes the formation of an ester bond between D-Ala and D-Lac. The resulting D-Ala-D-Lac depsipeptide replaces the D-Ala-D-Ala dipeptide in peptidoglycan synthesis, a substitution that decreases the affinity of the molecule for vancomycin and other glycopeptide antibiotic, teicoplanin, considerably [92, 93].

5.4. Resistance to other antibiotics

Since HA-MRSA strains are often MDR phenotype, drugs such as sulphonamides, tetracyclines, aminoglycosides, chloramphenicol and clindamycin were sidelined due to lack of activity, while vancomycin remained the mainstay of therapy. Resistance to sulphonamides and trimethoprim [94], tetracyclines [95–97], aminoglycosides [98–100], chloramphenicol [101] and clindamycin [102], occurring in *S. aureus* especially among MRSA was widely reported.

6. Therapeutic approach

Therapeutic approach to *S. aureus* infections depends on the type of infection, patient age, clinical manifestation of the disease, co-morbidity, antibacterial susceptibility of infecting organism and hospitalization. Various drugs as single agent and drug combinations have been used to treat *S. aureus* infection. In general, management of infections due to MRSA is difficult compared to that of MSSA. There are guidelines and reviews to help in the treatment of community and hospital infections of MRSA.

6.1. Topical anti-MRSA drugs

6.1.1. Mupirocin

Mupirocin is used as topical antibiotic to treat impetigo due to *S. aureus* and *S. pyogenes* [103]. The drug is also used for nasal decolonization of *S. aureus* [27]. Mupirocin belongs to monoxycarbolic acid class and it exerts antibacterial action by binding to isoleucyl t-RNA synthetase, thereby, inhibiting the protein synthesis [104]. The antibiotic shows excellent activity against Staphylococci and most Streptococci [105]. Clinical efficacy of mupirocin ointment in treating *S. aureus* superficial skin infections and wound infections was established [106–108]. Various reports also demonstrated effectiveness of mupirocin in nasal decolonization of *S. aureus* [25, 109, 110] that is a risk factor for MRSA infections in nosocomial settings.

6.1.2. Fusidic acid

Fusidic acid is an antibiotic, which belongs to a class referred to as fusidanes. Chemically it is a tetracyclic triterpenoid [111] and it binds to bacterial elongation factor G (EF-G), which results in impaired translocation process and inhibition of protein synthesis [112]. It has potent activity against *S. aureus* and clinically used in treatment of mild to moderately severe skin and soft-tissue infections, for example, impetigo, folicullitis, erythrasma, furunculosis, abscesses and infected traumatic wounds [113]. The efficacy of fusidic acid ointment in treatment of *S. aureus* infections is widely reported [114, 115]. The drug has also been used systemically to treat invasive *S. aureus* infections but its efficacy was questioned [116].

6.2. Systemic anti-MRSA drugs

6.2.1. Vancomycin

As discussed earlier, vancomycin remained the mainstay of therapy against MRSA infections in hospitalized patients for decades. Though the antibiotic was available for clinical use since 1958, it gained prominence among clinicians only after the surge in nosocomial MRSA infections in 1980s [73, 75]. Numerous reports documented the clinical efficacy of vancomycin in treating various MRSA infections in hospitalized patients [116–120]. The emergence and spread of hVISA and VISA strains has threatened the clinical utility of vancomycin. In addition, over the years, the mean MIC of vancomycin against susceptible MRSA

http://dx.doi.org/10.5772/67338

Newer-MRSA drug	Year of approval	Class	Source	Mode of action	Route of administration	References
Linezolid	2000	Oxazolidinone	Synthetic	Inhibition of protein synthesis	Oral & intra-venous	[126, 127]
Daptomycin	2003	Cyclic lipopeptide	*Streptomyces oseosporus*	Cell membrane depolarization	Intra-venous	[128, 129]
Tigecycline	2005	Glycylcyclines (Tetracyclines)	Semisynthetic	Inhibition of protein synthesis	Intra-venous	[130, 131]
Ceftaroline	2010	Cephalosporin (Beta-lactam)	Semisynthetic	Inhibition of cell wall synthesis	Intra-venous	[132, 133]
Telavancin	2013	Lipoglycopeptide	Semisynthetic	Inhibition of cell wall synthesis & cell membrane depolarization	Intra-venous	[134, 135]
Tedizolid	2014	Oxazolidinone	Synthetic	Inhibition of protein synthesis	Oral & intra-venous	[136, 137]
Dalbavancin	2014	Lipoglycopeptide	Semisynthetic	Inhibition of cell wall synthesis	Intra-venous	[138, 139]
Oritavancin	2014	Lipoglycopeptide	Semisynthetic	Inhibition of cell wall synthesis & cell membrane depolarization	Intra-venous	[140, 141]

Table 3. Newer anti-MRSA drugs.

populations has increased but within the susceptible range. This phenomenon is referred to as vancomycin MIC creep. There has been poor response to vancomycin therapy in patients infected with vancomycin-susceptible MRSA isolates which had vancomycin MIC at the higher end of susceptible range (2 mg/L) [121, 122]. Optimizing the dose regimen and drug delivery, in order to achieve the desired blood plasma concentration which would give the clinical efficacy is the way forward in preserving the clinical utility of vancomycin [91, 92].

6.2.2. Newer anti-MRSA drugs

The problem of MRSA infections in hospitals and lack of effective antibiotics other than vancomycin to treat them necessitated the discovery of novel anti-MRSA drugs. The continued efforts of researchers in discovering novel anti-MRSA drugs fructified resulting in arrival of number of newer anti-MRSA drugs for clinical use in the last 15 years [78, 123–125]. The following **Table 3** lists the newer anti-MRSA drugs that were approved by U.S. FDA for clinical use.

7. Alternative therapeutic approach

Apart from chemotherapeutic approach to tackle the *S. aureus* infection, alternatives such as agents which inhibit the virulent factors expression and vaccines have been investigated. Various phytochemical are also found to have anti-MRSA activity. All these are at investigational stages and more research is necessary to bring promising candidates for clinical usage.

7.1. Anti-virulence agents

Clinical use of agents which are not conventional antibiotics but able to inhibit the expression or function of the virulence factors, rendering the bacteria non-pathogenic is considered an alternative approach to tackle MRSA. Stripping microorganisms of their virulence properties without threatening their existence may offer a reduced selection pressure for drug-resistant mutations. Virulence-specific therapeutics would also avoid the undesirable dramatic alterations of the host microbiota that are associated with current antibiotics [142, 143].

Accessory gene regulator (*agr*)-mediated quorum sensing system of *S. aureus* plays a central role in pathogenesis of Staphylococci. Scientists identified small molecules which inhibited the *agr* system [144–146]. Active and passive immunization strategies targeting the virulence factors of *S. aureus* have also been explored [147].

7.2. Plants

Plants have immune system and other defensive mechanisms against microorganisms that cause plant diseases. Hence, the plants with huge diversity provide a vast source for exploration of anti-MRSA phytochemicals. *In vitro* Anti-MRSA activity of crude extracts of medicinal plants has been extensively reported [148]. Various phytochemicals such as β-asarone, Mansonone F, prenylated flavonoids and thymoquinone showed *in vitro* anti-MRSA activity [149–152].

Author details

Arumugam Gnanamani[1*], Periasamy Hariharan[2] and Maneesh Paul-Satyaseela[2,3]

*Address all correspondence to: gnanamani3@gmail.com

1 Microbiology Division, CSIR-CLRI, Adyar, Chennai, India

2 Orchid Pharma Ltd., Chennai, India

3 St Martha's Hospital, Bangalore, India

References

[1] Licitra G. Etymologia: Staphylococcus. Emerg Infect Dis. 2013;**19**:1553. DOI: 10.3201/eid1909.ET1909

[2] Greenwood D, O'Grady F. Scanning electron microscopy of *Staphylococcus aureus* exposed to some common anti-staphylococcal agents. J Gen Microbiol. 1972;**70**:263–270. DOI: 10.1099/00221287-70-2-263

[3] Foster T. Chapter 12: Staphylococcus. Medical Microbiology. 4th edition. Galveston (TX): University of Texas Medical Branch at Galveston, Galveston, Texas; 1996.

[4] Touhami A, Jericho MH, Beveridge TJ. Atomic force microscopy of cell growth and division in *Staphylococcus aureus*. J Bacteriol. 2004;**186**:3286–3295. DOI: 10.1128/JB.186.11.3286-3295.2004 DOI: 10.1128/JB.186.11.3286-3295.2004

[5] Liu GY, Essex A, Buchanan JT, Datta V, Hoffman HM, Bastian JF, Fierer J, Nizet V. *Staphylococcus aureus* golden pigment impairs neutrophil killing and promotes virulence through its antioxidant activity. J Exp Med. 2005;**202**:209–215. DOI: 10.1084/jem.20050846

[6] Blair JE. Factors determining the pathogenicity of staphylococci. Annu Rev Microbiol. 1958;**12**:491–506. DOI: 10.1146/annurev.mi.12.100158.002423

[7] Dinges MM, Orwin PM, Schlievert PM. Exotoxins of *Staphylococcus aureus*. Clin Microbiol Rev. 2000;**13**:16–34. DOI: 10.1128/CMR.13.1.16-34.2000

[8] Brown DF, Edwards DI, Hawkey PM, Morrison D, Ridgway GL, Towner KJ, Wren MW; Joint Working Party of the British Society for Antimicrobial Chemotherapy; Hospital Infection Society; Infection Control Nurses Association. Guidelines for the laboratory diagnosis and susceptibility testing of methicillin-resistant *Staphylococcus aureus* (MRSA). J Antimicrob Chemother. 2005;**56**:1000–1018. DOI: 10.1093/jac/dki372

[9] Archer GL. *Staphylococcus aureus*: a well-armed pathogen. Clin Infect Dis. 1998;**26**: 1179–1181. PMID: 9597249

[10] Tong SYC, Davis JS, Eichenberger E, Holland TL, Fowler Jr VG. *Staphylococcus aureus* infections: epidemiology, pathophysiology, clinical manifestations, and management. Clin Microbiol Rev. 2015;**28**:603–661. DOI: 10.1128/CMR.00134-14

[11] Kim HK, Falugi F, Missiakas DM, Schneewinda O. Peptidoglycan-linked protein A promotes T cell-dependent antibody expansion during *Staphylococcus aureus* infection. Proc Natl Acad Sci U S A. 2016;**113**:5718–5723. DOI: 10.1073/pnas.1524267113

[12] Gordon RJ, Lowy FD. Pathogenesis of methicillin-resistant *Staphylococcus aureus* infection. Clin Infect Dis. 2008;**46**:S350–S359. DOI: 10.1086/533591

[13] Vazquez V, Liang X, Horndahl JK, Ganesh VK, Smeds E, Foster TJ, and Hook M.. Fibrinogen is a ligand for the *Staphylococcus aureus* microbial surface components recognizing adhesive matrix molecules (MSCRAMM) bone sialoprotein-binding protein (Bbp). J Biol Chem. 2011;286:29797–29805. DOI: 10.1074/jbc.M110.214981

[14] Nilsson IM, Lee JC, Bremell T, Rydén C, Tarkowsk. The role of staphylococcal polysaccharide microcapsule expression in septicemia and septic arthritis. Infect Immun. 1997;**65**:4216–4221. PMCID: PMC175605

[15] Hong X, Qin J, Li T, Dai Y, Wang Y Liu Q, He L, Lu H, Gao Q, Lin Y, Li M. Staphylococcal Protein A promotes colonization and immune evasion of the epidemic healthcare-associated MRSA ST239. Front Microbiol. 2016;**7**:951. DOI: 10.3389/fmicb.2016.00951

[16] Voyich JM, Otto M, Mathema B, Braughton KR, Whitney AR, Welty D, Long RD, Dorward DW, Gardner DJ, Lina G, Kreiswirth BN, De Leo FR. Is panton-valentine leukocidin the major virulence determinant in community-associated methicillin-resistant *Staphylococcus aureus* disease? J Infect Dis. 2006;**194**:1761–1770. DOI: 10.1086/509506

[17] Genestier AL, Michallet MC, Prévost G, Bellot G, Chalabreysse L , Peyrol S, Thivolet F, Etienne J, Lina G, Vallette FM, Vandenesch F, Genestier L. *Staphylococcus aureus* Panton-Valentine leukocidin directly targets mitochondria and induces Bax-independent apoptosis of human neutrophils. J Clin Invest. 2005;**115**:3117–3127. DOI: 10.1172/JCI22684

[18] Bhakdi S, Tranum-Jensen J. Alpha-toxin of *Staphylococcus aureus*. Microbiol Rev. 1991;**55**:733–751. PMCID: PMC372845

[19] Postma B, Poppelier MJ, van Galen JC, Prossnitz ER, van Strijp JA, de Haas CJ, van Kessel KP. Chemotaxis inhibitory protein of *Staphylococcus aureus* binds specifically to the C5a and formylated peptide receptor. J Immunol. 2004;**172**:6994–7001. PMID: 15153520

[20] Edwards AM, Bowden MG, Brown EL, Laabei M, Massey RC. *Staphylococcus aureus* extracellular adherence protein triggers TNFα release, promoting attachment to endothelial cells via Protein A. PLoS One. 2012;**7**:e43046. DOI: 10.1371/journal.pone.0043046

[21] Argudin MA, Mendoza MC, Rodico MR. Food Poisoning and *Staphylococcus aureus* Enterotoxins. Toxins (Basel). 2010;**2**:1751–1773. DOI: 10.3390/toxins2071751

[22] Bukowski M, Wladyka B and Dubin G. Exfoliative Toxins of *Staphylococcus aureus*. Toxins (Basel). 2010;**2**:1148–1165. DOI: 10.3390/toxins2051148

[23] Kluytmans JA, Wertheim HF. Nasal carriage of *Staphylococcus aureus* and prevention of nosocomial infections. Infection. 2005;**33**:3–8. DOI: 10.1007/s15010-005-4012-9

[24] Williams REO. Healthy carriage of *Staphylococcus aureus*: its prevalence and importance. Bacteriol Rev. 1963;**27**:56–71. PMCID: PMC441169

[25] Wertheim HF, Verveer J, Boelens HA, van Belkum A, Verbrugh HA, Vos MC. Effect of mupirocin treatment on nasal, pharyngeal, and perineal carriage of *Staphylococcus aureus* in healthy adults. Antimicrob Agents Chemother. 2005;**49**:1465–1467. DOI: 10.1128/AAC.49.4.1465-1467.2005

[26] Perl TM. Prevention of *Staphylococcus aureus* infections among surgical patients: beyond traditional perioperative prophylaxis. Surgery. 2003;**134**:S10–S17. DOI: 10.1016/S0039

[27] Coates T, Bax R, Coates A. Nasal decolonization of *Staphylococcus aureus* with mupirocin: strengths, weaknesses and future prospects. J Antimicrob Chemother. 2009;**64**:9–15. DOI: 10.1093/jac/dkp159

[28] Kluytmans J. Nasal carriage of *Staphylococcus aureus*: epidemiology, underlying mechanisms, and associated risks. Clin Microbiol Rev. 1997;**10**:505–520. PMCID: PMC172932

[29] Fuda C, Suvorov M, Vakulenko SB, Mobashery S. The basis for resistance to β-lactam antibiotics by penicillin-binding protein 2a of methicillin-resistant *Staphylococcus aureus*. J Biol Chem. 2004;**279**:40802–40806. DOI: 10.1074/jbc.M403589200

[30] Rammelkamp CH, Maxon T. Resistance of *Staphylococcus aureus* to the action of penicillin. Exp Biol Med (Maywood). 1942;**51**:386–389. DOI: 10.3181/00379727-51-13986

[31] Kirby, WM. Extraction of highly potent penicillin inactivator from penicillin resistant Staphylocooci. Science. 1942;**99**:452–453. DOI: 10.1126/science.99.2579.452

[32] Bondi JA, Dietz CC. Penicillin resistant Staphylococci. Proc Royal Soc Exper Biol Med. 1945;**60**:55–58. PMID: 21004029

[33] Chambers HF. The changing epidemiology of *Staphylococcus aureus*? Emerg Infect Dis. 2001;**7**:178–182. DOI: 10.3201/eid0702.700178

[34] Lowy FD. Antimicrobial resistance: the example of *Staphylococcus aureus*. J Clin Invest. 2003;**111**:1265–1273. DOI: 10.1172/JCI18535

[35] Jevons PM. "Celbenin" – resistant Staphylococci. Br Med J. 1961;**1**:124–125. PMCID: PMC1952889

[36] Parker MT, Hewitt JH. Methicillin resistant *Staphylococcus aureus*. Lancet. 1970;**295**:800–804.

[37] Ayliffe GAJ. The progressive intercontinental spread of methicillin-resistant *Staphylococcus aureus*. Clin Infect Dis. 1997;**24**:S74–S79. PMID: 8994782

[38] Hartman A, Tomasz B. Altered penicillin-binding proteins in methicillin-resistant strains of *Staphylococcus aureus*. Antimicrob Agents Chemother. 1981;**19**:726–735. PMCID: PMC181513

[39] Wielders CLC, Fluit AC, Brisse S, Verhoef J, Schmitz FJ. mecA Gene is widely disseminated in *Staphylococcus aureus* population. J Clin Microbiol. 2002;**40**:3970–3975. PMCID: PMC139644

[40] Enright MC, Robinson DA, Randle G, Feil EJ, Grundmann H, Spratt BG. The evolutionary history of methicillin-resistant *Staphylococcus aureus* (MRSA). Proc Natl Acad Sci U S A. 2002;**99**:7687–7692. DOI: 10.1073/pnas.122108599

[41] Hiramatsu K, Katayama Y, Matsuo M, Sasaki T, Morimoto Y, Sekiguchi A, Baba T. Multidrug-resistant *Staphylococcus aureus* and future chemotherapy. J Infect Chemother. 2014;**20**:593–601. DOI: 10.1016/j.jiac.2014.08.001

[42] Ito T. Classification of Staphylococcal Cassette Chromosome *mec* (SCC*mec*): guidelines for reporting novel SCC*mec* elements. Antimicrob Agents Chemother. 2009;**53**(12):4961–4967. DOI: 10.1128/AAC.00579-09

[43] Ito T, Kuwahara-Arai K, Katayama Y, Uehara Y, Han X, Kondo Y, Hiramatsu K. Staphylococcal Cassette Chromosome mec (SCCmec) analysis of MRSA. Methods Mol Biol. 2014;**1085**:131–148. DOI: 10.1007/978-1-62703-664-1_8

[44] Naimi TS, LeDell KH, Como-Sabetti K, Borchardt SM, Boxrud DJ, Etienne J, Johnson SK, Vandenesch F, MD, Fridkin S, O'Boyle C, Danila RN, Lynfield R. Comparison of community- and health care-associated methicillin-resistant *Staphylococcus aureus* infection. JAMA. 2003;**290**:2976–2984. DOI: 10.1001/jama.290.22.2976

[45] Fridkin SK, Hageman JC, Morrison M, Sanza LT, Como-Sabetti K, Jernigan JA, Harriman K, Harrison LH, Lynfield R Farley MM. Methicillin-resistant *Staphylococcus aureus* disease in three communities. N Engl J Med. 2005;**352**:1436–1444. DOI: 10.1056/NEJMoa043252

[46] Hiramatsu K, Katayama Y, Yukawa H, Ito T. Molecular genetics of methicillin-resistant *Staphylococcus aureus*. Int J Med Microbiol. 2002;**292**:67–74. DOI: 10.1078/1438-4221-00192

[47] Hiramastsu K, Cui L, Kuroda M, Ito T. The emergence and evolution of methicillin-resistant *Staphylococcus aureus*. Trends Microbiol. 2001:**9**:486–493. PMID: 11597450

[48] Deresinski S. Methicillin-resistant *Staphylococcus aureus*: an evolutionary, epidemiologic, and therapeutic odyssey. Clin Infect Dis. 2005;**40**:562–573. DOI: 10.1086/427701

[49] Chambers HF, DeLeo FR. Waves of resistance: *Staphylococcus aureus* in the antibiotic era. Nat Rev Microbiol. 2009;**7**:629–641. DOI: 10.1038/nrmicro2200

[50] Uhlemann AC, Otto M, Lowy FD, DeLeoc FR. Evolution of community- and healthcare-associated methicillin-resistant *Staphylococcus aureus*. Infect Genet Evol. 2014;**21**:563–574. DOI: 10.1016/j.meegid.2013.04.030

[51] David MZ, Daum RS. Community-associated methicillin-resistant *Staphylococcus aureus*: epidemiology and clinical consequences of an emerging epidemic. Clin Microbiol Rev. 2010;**23**:616–687. DOI: 10.1128/CMR.00081-09

[52] Morrison MA, Hageman JC, Klevens RM. Case definition for community-associated methicillin-resistant *Staphylococcus aureus*. J Hosp Infect. 2006;**62**:241. DOI: 10.1016/j.jhin.2005.07.011

[53] DeLeo FR, Otto M, Kreiswirth BN, Chambers HF. Community-associated meticillin-resistant *Staphylococcus aureus*. Lancet. 2010;**375**:1557–1568. DOI: 10.1016/S0140-6736(09)61999-1

[54] King MD, Humphrey BJ, Wang YF, Kourbatova EV, Ray SM, Blumberg HM. Emergence of community-acquired methicillin-resistant *Staphylococcus aureus* USA 300 clone as the predominant cause of skin and soft-tissue infections. Ann Intern Med. 2006:**144**:309–317. PMID: 16520471

[55] Sowash MG, Uhlemann AC. Community-associated methicillin-resistant *Staphylococcus aureus* case studies. Methods Mol Biol. 2014;**1085**:25–69. DOI: 10.1007/978-1-62703-664-1_2

[56] Tokajian S. New epidemiology of *Staphylococcus aureus* infections in the Middle East. Clin Microbiol Infect. 2014;**20**:624–628. DOI: 10.1111/1469-0691.12691

[57] Hageman JC, Patel J, Franklin P, Miscavish K, McDougal L, Lonsway D, Khan FN. Occurrence of a USA300 vancomycin- intermediate *Staphylococcus aureus*. Diagn Microbiol Infect Dis. 2008;**62**:440–442. DOI: 10.1016/j.diagmicrobio.2008.08.005

[58] Mediavilla JR, Chen L, Mathema B., Kreiswirth BN. Global epidemiology of community-associated methicillin resistant *Staphylococcus aureus* (CA-MRSA). Curr Opin Microbiol. 2012;**15**(5):588–595. DOI: 10.1016/j.mib.2012.08.003

[59] Fletcher C. First clinical use of penicillin. Br Med J. 1984;**289**:1721–1723. PMCID: PMC1444782

[60] Chain E, Florey HW, Adelaide MB, Gardner AD, Heatley NG, Jennings MA, Orr-Ewing J, Sanders AG. Penicillin as a chemotherapeutic agent. Lancet. 1940;**ii**:226–228. DOI: 10.1016/S0140-6736(01)08728-1

[61] Klein E, Smith DL, Laxminarayanan R. Hospitalizations and deaths caused by methicillin-resistant *Staphylococcus aureus*, United States, 1999–2005. Emerg Infect Dis. 2007;**13**:1840–1846. DOI: 10.3201/eid1312.070629

[62] Köck R, Becker K, Cookson B, van Gemert-Pijnen JE, Harbarth S, Kluytmans J, Mielke M, Peters G, Skov RL, Struelens MJ, Tacconelli E, Navarro Torné A, Witte W, Friedrich AW. Methicillin-resistant *Staphylococcus aureus* (MRSA): burden of disease and control challenges in Europe. Euro Surveill. 2010;**14**;15:19688. PMID: 20961515

[63] Emmerson AM, Jones AM.. The quinolones: decades of development and use. J Antimicrob Chemother. 2003;**51**:13–20. DOI: 10.1093/jac/dkg208

[64] King DE, Malone R, Lilley SH. New classification and update on the quinolone antibiotics. Am Fam Physician 2000;**61**:2741–2748. PMID: 10821154

[65] Ball P. Quinolone generations: natural history or natural selection? J Antimicrob Chemother. 2000;**46**:17–24. PMID: 10997595

[66] Hooper DC. Mode of action of fluoroquinolones. Drugs. 1999;**58**:6–10. PMID: 10553698

[67] Hooper DC. Mechanisms of action and resistance of older and newer fluoroquinolones. Clin Infect Dis. 2000;**31**:S24–S28. DOI: 10.1086/314056

[68] Raviglione MC, Boyle JF, Mariuz P, Pablos-Mendez A, Cortes H, Merlo A. Ciprofloxacin-resistant methicillin-resistant *Staphylococcus aureus* in an acute-care hospital. Antimicrob Agents Chemother. 1990;**34**:2050–2054. PMCID: PMC171997

[69] Weber SG, Gold HS, Hooper DC, Karchmer AW, Carmeli Y. Fluoroquinolones and the risk for methicillin-resistant *Staphylococcus aureus* in hospitalized patients. Emerg Infect Diseases. 2003;**9**:1415–1422. DOI: 10.3201/eid0911.030284

[70] Dalhoff A, Schubert S. Dichotomous selection of high-level oxacillin resistance in *Staphylococcus aureus* by fluoroquinolones. Int J Antimicrob Agents. 2010;**36**:216–221. DOI: 10.1016/j.ijantimicag.2010.04.014

[71] Jones RN, Mendes RE, Sader H. Ceftaroline activity against pathogens associated with complicated skin and skin structure infections: results from an international surveillance study. J Antimicrob Chemother. 2010;**65**:iv17–iv31. DOI: 10.1093/jac/dkq252

[72] Dalhoff A. Global fluoroquinolone resistance epidemiology and implications for clinical use. Interdiscip Perspect Infect Dis. 2012;**2012**:976273. DOI: 10.1155/2012/976273

[73] Levine DP. Vancomycin: a history. Clin Infect Dis. 2006;**42**:S5–S12. DOI: 10.1086/491709

[74] Kirst HA. Historical yearly usage of vancomycin. Antimicrob Agents Chemother. 1998;**42**:1303–1304. PMCID: PMC105816

[75] Moellering Jr RC. Vancomycin: a 50-year reassessment. Clin Infect Dis. 2005;**42**(Suppl. 1):S3–S4. DOI: 10.1086/491708

[76] Rodvold KA, McConeghy KW. Methicillin resistant *Staphylococcus aureus* therapy: past, present and future. Clin Infect Dis. 2014;**58**:S20–S27. DOI: 10.1093/cid/cit614

[77] Courvalin P. Vancomycin resistance in Gram-positive cocci. Clin Infect Dis. 2006;**42**:S25–S34. DOI: 10.1086/491711

[78] Howden BP, Davies JK, Johnson PDR, Stinear TP, Grayson ML. Reduced vancomycin susceptibility in *Staphylococcus aureus*, including vancomycin-intermediate and heterogeneous vancomycin intermediate Strains: resistance mechanisms, laboratory, detection, and clinical implications. Clin Microbiol Rev. 2010;**23**:99–139. DOI: 10.1128/CMR.00042-09

[79] Hiramatsu K, Hanaki H, Ino T, Yabuta K, Oguri T, Tenover FC. Methicillin-resistant *Staphylococcus aureus* clinical strain with reduced vancomycin susceptibility. J Antimicrob Chemother. 1997;**40**:135–136. PMID: 9249217

[80] Appelbaum PC. Reduced glycopeptide susceptibility in methicillin-resistant *Staphylococcus aureus* (MRSA). Int J Antimicrob Agents. 2007;**30**:398–408. DOI: 10.1016/j.ijantimicag.2007.07.011

[81] Appelbaum PC. The emergence of vancomycin-intermediate and vancomycin-resistant *Staphylococcusaureus*.ClinMicrobiolInfect.2006;**12**:16–23.DOI:10.1111/j.1469-0691.2006.01344.x

[82] Cui L, Ma X, Sato K, Okuma K, Tenover FC, Mamizuka EM, Gemmell CG, Kim MN, Ploy MC, El-Solh N, Ferraz V, Hiramatsu K. Cell wall thickening is a common feature of vancomycin resistance in *Staphylococcus aureus*. J Clin Microbiol. 2003;**41**:5–14. PMCID: PMC149586

[83] Gardete S and Tomasz A. Mechanisms of vancomycin resistance in *Staphylococcus aureus*. J Clin Invest. 2014;**124**:2836–2840. PMCID: PMC149586

[84] Satola SW, Farley MM, Anderson KF, Patel JB. Comparison of detection methods for heteroresistant vancomycin-intermediate *Staphylococcus aureus*, with the population analysis profile method as the reference method. J Clin Microbiol. 2011;**49**:177–183. DOI: 10.1128/JCM.01128-10

[85] Riederer K, Shemes S, Chase P, Musta A, Mar A. Detection of intermediately vancomycin-susceptible and heterogeneous *Staphylococcus aureus* isolates: comparison of Etest and Agar screening methods. J Clin Microbiol. 2011;**49**(6):2147–2150. DOI: 10.1128/JCM.01435-10

[86] Ford BA. Identification of low-level vancomycin resistance in *Staphylococcus aureus* in the era of informatics. J Clin Microbiol. 2016;**54**:836–839. DOI: 10.1128/JCM.00071-16

[87] Zhang S, Sun X, Chang W, Dai Y, Ma X. Systematic review and meta-analysis of the epidemiology of vancomycin-intermediate and heterogeneous vancomycin-intermediate *Staphylococcus aureus* isolates. PLoS One. 2015;**10**:e0136082. DOI: 10.1371/journal.pone.0136082

[88] Pai MP, Neely M, Rodvold KA, Lodise TP. Innovative approaches to optimizing the delivery of vancomycin in individual patients. Adv Drug Deliv Rev. 2014;**77**:50–57. DOI: 10.1016/j.addr.2014.05.016

[89] Álvarez R, López Cortés LE, Molina J, Cisneros JM, Pachón J. Optimizing the clinical use of vancomycin. Antimicrob Agents Chemother. 2016;**60**:2601–2609. DOI: 10.1128/AAC.03147-14

[90] Sievert DM *et al.*, CDC. *Staphylococcus aureus* Resistant to vancomycin – United States, MMWR. 2002;**51**:565–567. PMID: 12139181

[91] Sievert DM, Rudrik JT, Patel JB, McDonald LC, Wilkins MJ, Hageman JC. Vancomycin-resistant *Staphylococcus aureus* in the United States, 2002–2006. Clin Infect Dis. 2008;**46**:668–674. DOI: 10.1086/527392

[92] Depardieu F, Podglajen I, Leclercq R, Collatz E, Courvalin P. Modes and modulations of antibiotic resistance gene expression. Clin Microbiol Rev. 2007;**20**:79–114. DOI: 10.1128/CMR.00015-06

[93] Périchon B, Courvalin P. VanA-type Vancomycin-resistant *Staphylococcus aureus*▿. Antimicrob Agents Chemother. 2009;**53**:4580–4587. DOI: 10.1128/AAC.00346-09

[94] Then RL, Kohl I, Burdeska A.. Frequency and transferability of trimethoprim and sulfonamide resistance in methicillin-resistant *Staphylococcus aureus* and *Staphylococcus epidermidis*. J Chemother. 1992;4(2):67–71. PMID: 1629749

[95] Trzcinski K, Cooper BS, Hryniewicz W, Dowson CG. Expression of resistance to tetracyclines in strains of methicillin-resistant *Staphylococcus aureus*. J Antimicrob Chemother. 2000;**45**(6):763–770. PMID: 10837427

[96] Schmitz FJ, Krey A, Sadurski R, Verhoef J, Milatovic D, Fluit AC; European SENTRY Participants. Resistance to tetracycline and distribution of tetracycline resistance genes in European *Staphylococcus aureus* isolates. J Antimicrob Chemother. 2001;**47**:239–240. DOI: 10.1093/jac/47.2.239

[97] Fluit AC, Florijn A, Verhoef J, Milatovic D. Presence of tetracycline resistance determinants and susceptibility to tigecycline and minocycline. Antimicrob Agents Chemother. 2005;**49**:1636–1638. DOI: 10.1128/AAC.49.4.1636-1638.2005

[98] Storrs MJ, Courvalin P, Foster TJ. Genetic analysis of gentamicin resistance in methicillin- and gentamicin-resistant strains of *Staphylococcus aureus* isolated in Dublin hospitals. Antimicrob Agents Chemother. 1988;**32**:1174–1181. PMCID: PMC172372

[99] Freitas FI, Guedes-Stehling E, Siqueira-Júnior JP. Resistance to gentamicin and related aminoglycosides in *Staphylococcus aureus* isolated in Brazil. Lett Appl Microbiol. 1999;**29**:197–201. PMID: 10530041

[100] Schmitz FJ, Fluit AC, Gondolf M, Beyrau R, Lindenlauf E, Verhoef J, Heinz HP, Jones ME. The prevalence of aminoglycoside resistance and corresponding resistance genes in clinical isolates of staphylococci from 19 European hospitals. J Antimicrob Chemother. 1999;**43**:253–259. PMID: 11252331

[101] Fayyaz M, Mirza IA, Ahmed Z, Abbasi SA, Hussain A, Ali S. In vitro susceptibility of chloramphenicol against methicillin-resistant *Staphylococcus aureus*. J Coll Physicians Surg Pak. 2013;**23**:637–640. DOI: 09.2013/JCPSP.637640

[102] Frank AL, Marcinak JF, Mangat PD, Tjhio JT, Kelkar S, Schreckenberger PC, Quinn JP. Clindamycin treatment of methicillin-resistant *Staphylococcus aureus* infections in children. Pediatr Infect Dis J. 2002;**21**:530–534. PMID: 12182377

[103] Putnam CD, Reynolds MS. Mupirocin: a new topical therapy for impetigo. J Pediatr Health Care. 1989:**3**:224–227. PMID: 2502615

[104] Parenti MA, Hatfield SM, Leyden JJ. Mupirocin: a topical antibiotic with a unique structure and mechanism of action. Clin Pharm. **1987**;6:761–770. PMID: 3146455

[105] Ward A, Campoli-Richards DM. Mupirocin. A review of its antibacterial activity, pharmacokinetic properties and therapeutic use. Drugs.1986;**32**:425–444. PMID: 3098541

[106] Villiger JW, Robertson WD, Kanji K, Ah Chan M, Fetherston J, Hague IK, Haycock D, Hunter P. A comparison of the new topical antibiotic mupirocin ('Bactroban') with oral

antibiotics in the treatment of skin infections in general practice. Curr Med Res Opin. 1986;**10**:339–345. DOI: 10.1185/03007998609111100

[107] Bork K, Brauers J, Kresken M. Efficacy and safety of 2% mupirocin ointment in the treatment of primary and secondary skin infections--an open multicentre trial. Br J Clin Pract. 1989;**43**:284–288. PMID: 2516463

[108] Rode H, Hanslo D, de Wet PM, Millar AJ, Cywes S. Efficacy of mupirocin in methicillin-resistant *Staphylococcus aureus* burn wound infection. Antimicrob Agents Chemother. 1989;**33**:1358–1361. PMCID: PMC172654

[109] Gaspar MC, Uribe P, Sánchez P, Coello R, Cruzet F. Hospital personnel who are nasal carriers of methicillin-resistant *Staphylococcus aureus*. Usefulness of treatment with mupirocin. Enferm Infecc Microbiol Clin. 1992;**10**:107–110. PMID: 1643130.

[110] van Rijen M, Bonten M, Wenzel R, Kluytmans J. Mupirocin ointment for preventing *Staphylococcus aureus* infections in nasal carriers. Cochrane Database Syst Rev. 2008;**8**:CD006216. DOI: 10.1002/14651858.CD006216.pub2.

[111] Godtfredsen W, Roholt K, Tybring L. Fucidin: a new orally active antibiotic. *Lancet*. 1962;**1**:928–931. PMID: 13899434

[112] Dobie D, Gray J. Fusidic acid resistance in *Staphylococcus aureus*. Arch Dis Child. 2004;**89**:74–77. DOI: 10.1136/adc.2003.019695

[113] Wilkinson JD. Fusidic acid in dermatology. Br J Dermatol. 1998;**139**:37–40. PMID: 9990411

[114] Morley PA, Munot LD. A comparison of sodium fusidate ointment and mupirocin ointment in superficial skin sepsis. Curr Med Res Opin. 1988;**11**:142–148. DOI: 10.1185/03007998809110457.

[115] White DG, Collins PO, Rowsell RB. Topical antibiotics in the treatment of superficial skin infections in general practice--a comparison of mupirocin with sodium fusidate. J Infect. 1989;**18**:221–229. PMID: 2501394

[116] Wood MJ. The comparative efficacy and safety of teicoplanin and vancomycin. J Antimicrob Chemother. 1996;**37**:209–222. DOI: 10.1093/jac/37.2.209

[117] Svetitsky S, Leibovici L, Paul M. Comparative efficacy and safety of vancomycin versus teicoplanin: systematic review and meta-analysis. Antimicrob Agents Chemother. 2009;**53**:4069–4079. DOI: 10.1128/AAC.00341-09

[118] Dodds TJ, Hawke CI. Linezolid versus vancomycin for MRSA skin and soft tissue infections (systematic review and meta-analysis). ANZ J Surg. 2009;**79**:629–635. DOI: 10.1111/j.1445–2197.2009.05018.x

[119] Kalil AC, Murthy MH, Hermsen ED, Neto FK, Sun J, Rupp ME. Linezolid versus vancomycin or teicoplanin for nosocomial pneumonia: a systematic review and meta-analysis. Crit Care Med. 2010;**38**:1802–1808. DOI: 10.1097/CCM.0b013e3181eb3b96

[120] Eckmann C, Dryden M. Treatment of complicated skin and soft-tissue infections caused by resistant bacteria: value of linezolid, tigecycline, daptomycin and vancomycin. Eur J Med Res. 2010:15:554–563. DOI: 10.1186/2047-783X-15-12-554

[121] Deresinski S. Counterpoint: vancomycin and *Staphylococcus aureus* – an antibiotic enters obsolescence. Clin Infect Dis. 2007;**44**:1543–1548. DOI: 10.1086/518452

[122] Dhand A, Sakoulas G. Reduced vancomycin susceptibility among clinical *Staphylococcus aureus* isolates ('the MIC Creep'): implications for therapy. F1000 Med Rep. 2012;**4**:4. DOI: 10.3410/M4-4

[123] Micek ST. Alternatives to vancomycin for the treatment of methicillin-resistant *Staphylococcus aureus* Infections. Clin Infect Dis. 2007;**45**:S184–S190. DOI: 10.1086/519471.

[124] Ohlsen K. Novel Antibiotics for the Treatment of *Staphylococcus aureus*. Expert Rev Clin Pharmacol. 2009;**2**:661–672. DOI: 10.1586/ecp.09.26

[125] Kurosu M, Siricilla S, Mitachi K. Advances in MRSA drug discovery: where are we and where do we need to be? Expert Opin Drug Discov. 2013;**8**:1095–1116. DOI: 10.1517/17460441.2013.807246

[126] Stevens DL, Dotter B, Madaras-Kelly K. A review of linezolid: the first oxazolidinone antibiotic. Expert Rev Anti Infect Ther. 2004;**2**:51–59. PMID: 15482171

[127] Watkins RR, Lemonovich TL, File TM. An evidence-based review of linezolid for the treatment of methicillin-resistant *Staphylococcus aureus* (MRSA): place in therapy. Core Evidence. 2012;**7**:131–143. DOI: 10.2147/CE.S33430

[128] Alder JD. Daptomycin: a new drug class for the treatment of Gram-positive infections. Drugs Today (Barc). 2005;**41**:81–90. DOI: 10.1358/dot.2005.41.2.882660

[129] Steenbergen JN, Alder J, Thorne GM, Tally FP. Daptomycin: a lipopeptide antibiotic for the treatment of serious Gram-positive infections. J Antimicrob Chemother. 2005;**55**:283–288. DOI: 10.1093/jac/dkh546

[130] Doan TL, Fung HB, Mehta D, Riska PF. Tigecycline: a glycylcycline antimicrobial agent. Clin Ther. 2006;**28**:1079–1106. DOI: 10.1016/j.clinthera.2006.08.011

[131] Stein GE, Craig WA. Tigecycline: a critical analysis. Clin Infect Dis. 2006:**43**:518–524. DOI: 10.1086/505494

[132] Laudano JB. Ceftaroline fosamil: a new broad-spectrum cephalosporin. J Antimicrob Chemother. 2011:**66**:iii11–iii18. DOI: 10.1093/jac/dkr095

[133] Shirley DA, Heil EL, Johnson JK. Ceftaroline fosamil: a brief clinical review. Infect Dis Ther. 2013;**2**:95–110. DOI: 10.1007/s40121-013-0010-x

[134] Scott LJ. Telavancin: a review of its use in patients with nosocomial pneumonia. Drugs. 2013;**73**:1829–1839. DOI: 10.1007/s40265-013-0144-x.

[135] Sandrock CE, Shorr AF.The role of telavancin in hospital-acquired pneumonia and ventilator-associated pneumonia. Clin Infect Dis. 2015;**61**:S79–S86. DOI: 10.1093/cid/civ535

[136] Wong E, Rab S. Tedizolid phosphate (sivextro): a second-generation oxazolidinone to treat acute bacterial skin and skin structure infections. P T. 2014:**39**:555–579. DOI: PMCID: PMC4123804

[137] Rybak JM, Roberts K.. Tedizolid Phosphate: a next-generation oxazolidinone. Infect Dis Ther. 2015;**4**:1–14. DOI: 10.1007/s40121-015-0060-3

[138] Juul JJ, Mullins CF, Peppard WJ, Huang AM. New developments in the treatment of acute bacterial skin and skin structure infections: considerations for the effective use of dalbavancin. Ther Clin Risk Manag. 2016;**12**:225–232. DOI: 10.2147/TCRM.S71855

[139] Leuthner KD, Buechler KA, Kogan D, Saguros A, Lee HS. Clinical efficacy of dalbavancin for the treatment of acute bacterial skin and skin structure infections (ABSSSI). Ther Clin Risk Manag. 2016;**12**:931–940. DOI: 10.2147/TCRM.S86330

[140] Markham A. Oritavancin: first global approval. Drugs. 2014;**74**:1823–1828. DOI: 10.1007/s40265-014-0295-4

[141] Mattox J, Belliveau P, Durand C. Oritavancin: a novel lipoglycopeptide. Consult Pharm. 2016;**31**(2):86–95. DOI: 10.4140/TCP.n.2016.86

[142] Cegelski L, Marshall GR, Eldridge GR, Hultgren SJ. The biology and future prospects of antivirulence therapies. Nat. Rev. Microbiol. 2008;**6**:17–27. DOI: 10.1038/nrmicro1818

[143] Rasko DA, Sperandio V. Anti-virulence strategies to combat bacteria-mediated disease. Nat Rev Drug Discov. 2010;**9**:117–128. DOI: 10.1038/nrd3013

[144] Queck SY, Jameson-Lee M, Villaruz AE, Bach TH, Khan BA, Sturdevant DE, Ricklefs SM, Li M, Otto M.. RNAIII-independent target gene control by the agr quorum-sensing system: insight into the evolution of virulence regulation in *Staphylococcus aureus*. Mol Cell. 2008;**32**:150–158. DOI: 10.1016/j.molcel.2008.08.005

[145] Singh R, Ray P. Quorum sensing-mediated regulation of staphylococcal virulence and antibiotic resistance. Future Microbiol. 2014;**9**:669–681. DOI: 10.2217/fmb.14.31.

[146] Sully EK, Malachowa N, Elmore BO, Alexander SM, Femling JK, Gray BM, DeLeo FR, Otto M, Cheung AL, Edwards BS, Sklar LA, Horswill AR, Hall PR, Gresham HD. Selective chemical inhibition of agr quorum sensing in *Staphylococcus aureus* promotes host defense with minimal impact on resistance. PLoS Pathogens. 2014;**10**:e1004174. 10.1371/journal.ppat.1004174

[147] Giersing BK, Dastghey SS, Modjarrad K, Moorthy V. Status of vaccine research and development of vaccines for *Staphylococcus aureus*. Vaccine. 2016;**34**:2962–2966. DOI: 10.1016/j.vaccine.2016.03.110

[148] Kali A. Antibiotics and bioactive natural products in treatment of methicillin resistant *Staphylococcus aureus*: a brief review. Pharmacogn Rev. 2015;**9**:29–34. DOI: 10.4103/0973-7847.156329

[149] Sujina I, Prabhu V, Hemlal H, Ravi S. Essential oil composition, isolation of β-asarone and its antibacterial and MRSA activity from the rhizome of *Acorus calamus*. J Pharm Res. 2012;**5**:3437–3440.

[150] Shin DY, Kim HS, Min KH, Hyun SS, Kim SA, Huh H, Choi EC, Choi YH, Kim J, Choi SH, Kim WB, Suh YG. Isolation of a potent anti-MRSA sesquiterpenoid quinone from *Ulmus davidiana* var. *japonica*. Chem Pharm Bull (Tokyo). 2000;**48**:1805–1806. PMID: 11086922

[151] Sasaki H, Kashiwada Y, Shibata H, Takaishi Y. Prenylated flavonoids from Desmodium caudatum and evaluation of their anti-MRSA activity. Phytochemistry. 2012;**82**:136–142. DOI: 10.1016/j.phytochem.2012.06.007

[152] Hariharan P, Paul-Satyaseela M, Gnanamani A. In vitro profiling of antimethicillin-resistant *Staphylococcus aureus* activity of thymoquinone against selected type and clinical strains. Lett Appl Microbiol. 2016;**62**:283–289. DOI: 10.1111/lam.12544

Section 2

Transmission of *Staphylococcus aureus*

Chapter 2

Role of Nasal *Staphylococcus aureus* Carriage in Transmission Among Contact Athletes

Kotaro Suzuki

Additional information is available at the end of the chapter

http://dx.doi.org/10.5772/66936

Abstract

This chapter focuses on *Staphylococcus aureus* (SA) infections in athletes. Previous SA infection studies performed starting in the 1980s examined close physical contact athletes, with a focus primarily on injured skin. However, more recent studies of skin SA transmission in athletes were conducted using molecular epidemiology. When participants in sports having a greater duration of competition were examined, results indicated that there was prolonged contact between athletes on the same team and athletes from other teams. These findings demonstrate that effective measures for preventing SA infections are urgently needed. Factors that can affect skin SA infections include high rates of SA nasal colonization, the type of "position on a team," repeated skin-to-skin contact, and perspiration that occurs during exercise in SA nasal carriers. Thus, it should be possible to utilize molecular typing methods to assess skin-to-skin contact in athletes. This study summarizes the current understanding of SA infections in athletes. In order to develop preventive strategies, it will be necessary to further elucidate the predisposing factors and mechanisms behind SA infections and the subsequent transmission in athletes.

Keywords: athletes, transmission, physical contact, genotyping

1. Introduction

Staphylococcus aureus (SA) infections and its transmission among athletes have long been of interest to sports medicine scientists. SA is very well adapted to colonize the human skin, as the human body provides major ecological niches for the species. Although originally thought to be a nosocomial pathogen, it has become a rapidly emerging, problematic infection in athletes [1]. When outbreaks occur, the infection is spread through repeated skin-to-skin contact, especially due to physical contact between the broken skin of players during games and practices. In addition, sharing contaminated equipment [2], turf burns, and shaving [3] also contribute to the high incidence of infection among athletes. A review of past studies of SA infections in athletes suggested that the risk factors associated with the outbreaks could be classified into three categories (**Figure 1**). These include direct contact (which refers to "contact sports" where physical contact between players is an acceptable part of the sport), nosocomial infections, and skin wounds. The infections that occur during these outbreaks can also disrupt or potentially eliminate the opportunity for a team to compete at the highest level of their sport. Furthermore, outbreaks of infectious diseases can additionally spread to the player's social contacts and propagate within their communities [4, 5]. However, the association between direct physical contact and the SA transmission has yet to be fully understood.

The primary goal of the review presented in this chapter is to provide a better understanding of SA infections and the potential relationship with the associated sports activity. This chapter is divided into four parts, with the first section summarizing the latest insights into the sports activity related to SA infections and risk factors. In the second part, we focus on the latest insights into the determinants of SA nasal carriage and skin infections. As nasal carriers may be the reservoir responsible for the transmission in athletes and teams, this section describes the first high-throughput SA nasal carriage effort for large numbers of SA from athletes. As nasal carriers may be the reservoir responsible for the transmission in athletes and teams, this section describes the first high-throughput SA nasal carriage genotyping effort that has been used to examine the SA transmission in athletes. Previous reports have shown that nasal carriage may play a key role in the epidemiology and pathogenesis of SA infections [6, 7]. The third part presents information on our current understanding on how SA can thrive on the skin and be easily transmitted from person to person via sweat. Thus, when a sport involves physical contact, this route is likely to be the major mode of transmission between the athletes. In the final section, we discuss the high-throughput genotyping effort that has been undertaken in order to investigate SA transmission in athletes.

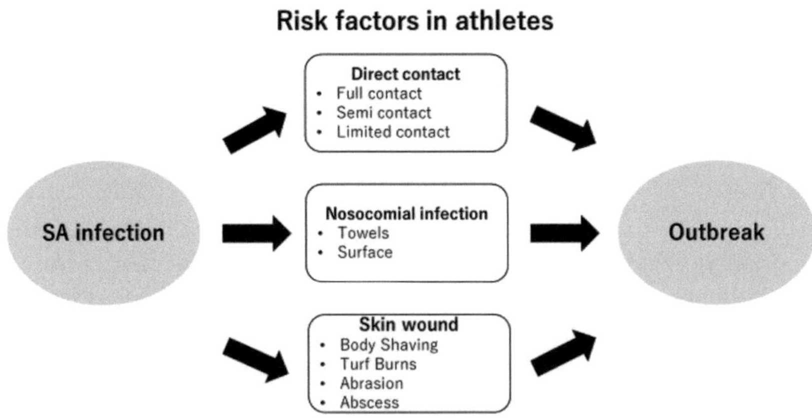

Figure 1. A hypothetical example of a *Staphylococcus aureus* infection resulting in an outbreak within an athletic setting. *Staphylococcus aureus* outbreaks are classified into three categories, which include "direct contact," "nosocomial infection," and "skin wound." These hypothetical schemes will need to be further examined in future experiments.

http://dx.doi.org/10.5772/66936

The discovery of SA outbreaks in athletes taking part in physical contact sports is not new. In 1982, Bartlett et al. published the first scientific paper on SA infections in athletes [8]. This study examined 26 players of a high school football team and reported finding a total of 55 lesions, with two players found to have Methicillin-Susceptible Staphylococcus Aureus (MSSA), while 24 had methicillin-resistant SA (MRSA). There was no pathogen growth observed for any of the players. There were three essential findings observed and confirmed by the authors in subsequent studies. First, the majority of the lesions observed were located on the extremities and in areas not usually covered by a football uniform or other apparels. Second, 61% of the affected players reported the development of a furuncle at the site of a previously open wound, while 27% reported the development of a furuncle at the site of a previous bruise. Third, cultures obtained from the lesions of two players grew SA that was sensitive to nafcillin, clindamycin, erythromycin, cephalosporin, tetracycline, and sulfa and resistant to penicillin and ampicillin. This is of importance, as infections with drug-resistant bacteria may lead to longer and more costly hospital care, in addition to an increase in the risk of dying from the infection.

Since this initial report, various infectious disease outbreaks have been reported [9]. Sosin et al. [10] additionally reported an outbreak of furuncles in athletes in the state of Kentucky in the USA. The outbreak involved members of the high school football and basketball teams, with a total of 62 lesions reported in these affected athletes. In this school, the basketball season overlapped with the end of the football season, with the two teams sharing a locker room. In addition, six of the players participated on both of the teams. Based on these findings, the authors hypothesized that close physical contact was a risk factor for SA transmission in athletes. The majority of affected players were treated with oral antibiotics, with three players developing infections that did not respond to oral therapy, thereby requiring hospitalization for intravenous antibiotic therapy. One of these hospitalized players subsequently developed a disseminated SA infection and a lung abscess. A total of 81% of the observed lesions were found on the extremities. Moreover, players who sustained a skin injury were three times more likely to develop an infection compared to those who did not report any skin injury. The use of the school showers and locker room and the sharing of clothing and towels were not found to be risk factors for SA infections. Although SA was isolated from 14 of 52 (27%) nasal cultures collected from the athletes, this was not higher than the proportion of SA-positive nasal cultures found in a group of student controls.

These studies provided valuable information and have helped encourage the development of subsequent investigations into additional components affecting SA skin infections. Furthermore, the majority of all of the studies performed because the initial publication has focused on investigations of the effects of physical contact on SA skin infection.

As Suzuki and Tagami [11] discussed in detail, physical contact contributes to SA transmission. Their study identified several factors including an outbreak of SA skin infection in a collegiate men's rugby team. The athletes examined had all started the rugby season with a training camp that was in close proximity to where they lived and which was conducted between August 4 and 25 in 2011 (**Figure 2A**). SA infections were found in 14 (20%) of the 69 healthy rugby players between September 10 and October 21 of 2011 (**Table 1**). One team member required hospitalization during October 2011 in order to treat an abscess that was secondary to the SA infection. As other members of the team also developed skin infections,

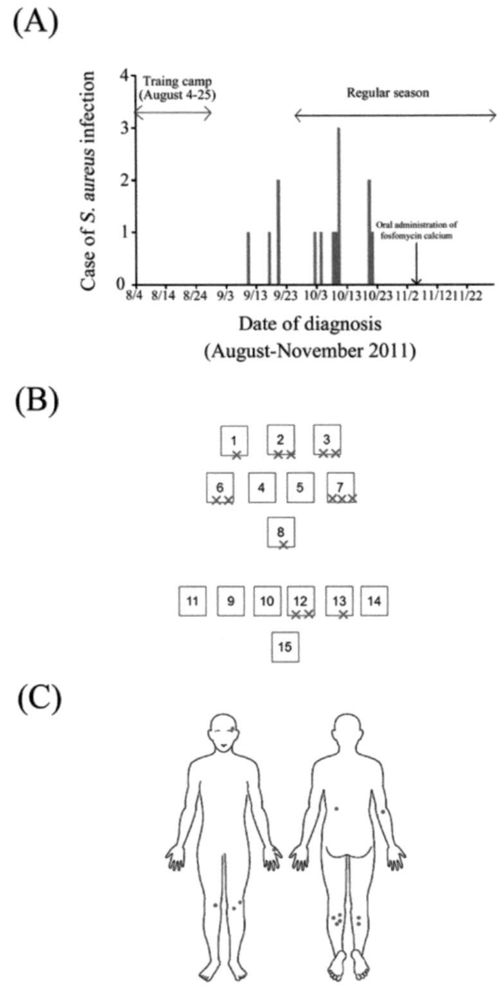

Figure 2. Epidemic curve of the initial skin infections due to *Staphylococcus aureus* among collegiate rugby players (A). Field position diagram of players who developed *Staphylococcus aureus* infection (B). See **Table 1** for position-specific attack rates. 1 and 3, prop; 2, hooker; 4 and 5, lock; 6 and 7, flanker; 8, number 8; 9, scrum half; 10, fly half; 11 and 14, wings; 12 and 13, center back; and 15, fullback. Players infected with *Staphylococcus aureus* (×). Anatomical locations of infection sites (C). Front and back side of body. Filled circles, skin lesions (unpublished data).

screening was begun on October 6. The dispersion of the outbreak was trimodal, with 28, 50, and 21% occurring between September 10 and 20, between October 2 and 10, and between October 20 and 21 (**Figure 2A**). The infections developed in 11 forwards and three back-positioned players (**Figure 2B**). The infection rate was higher among the forwards versus the back positions (28 vs. 7%) (**Table 1**), with the flanker position exhibiting a greater likelihood of becoming infected compared to the other players. The forehead, back, elbow, and thumb comprised the primary sites of SA infection (**Figure 2B**). Infections tended to occur most frequently on the areas that were not covered by athletic apparel, such as the elbows, forearms, knees, and lower legs [1, 8]. These trends suggest that competitive practices lead to repeated direct contact. Nine crural abscesses were located on the front and back of the legs near or on

http://dx.doi.org/10.5772/66936

Position		No. of infected players (%) $n = 14$	Total no. on team (%) $n = 69$	Position-specific attack rate (%)*
Forwards	Props	3 (21)	7 (10)	28
	Hooker	2 (14)	5 (7)	
	Locks	0	10 (14)	
	Flankers	5 (35)	14 (20)	
	Number 8	1 (7)	2 (2.8)	
Backs	Scrum half	0	6 (8.6)	7
	Fly half	0	4 (5.7)	
	Wings	0	5 (7.2)	
	Fullback	0	7 (9.8)	
	Center back	3 (21)	9 (13)	

*Attack rate = no. of infected players/total no. on team, per position.
The overall attack rate = 10%.

Table 1. Position-specific attack rates of clinical and *Staphylococcus aureus* and soft tissue infections among members of a rugby team (unpublished data).

the knee. In addition, most of the lesions were located on the extremities in areas that are not usually covered by the rugby uniform. However, SA isolates that resulted from the sharing of items, such as the contact bag, tackle bag, and bibs, were limited. Starting on November 3, oral fosfomycin calcium (1 g, two tablets) was administered three times daily for 5 days. After 22 days, there was a decrease in the number of SA nasal carriers. This intervention prevented further dissemination of the SA infection among the team members.

This report shows that the epidemic curves can provide considerable information about the outbreaks such as the pattern of the spread, magnitude, time trend, and exposure and disease incubation periods. The epidemic curve in the present study was trimodal, with continuous, intermittent exposure and gradual increases in the numbers of infections. This type of epidemic curve is typical of person-to-person spread [12]. Classic epidemic curves from propagating outbreaks comprise successively taller peaks, distanced one incubation period apart. The two most common sources of SA spread were contaminated hands and physical contact with athletes [13, 14]. Multimodal peaks are representative of SA outbreaks and comprise a risk factor for such outbreaks during physical contact sports [8]. These previous studies support a person-to-person contact method for the transmittal of the disease, with skin injuries serving as an entrance point for the infectious organisms.

2. When do *S. aureus* outbreaks occur in athletes?

Outbreaks of SA infections in an athletic setting have also occurred during the regular season (with the term "regular season" referring to the sport's league competitive period) [15]. During the regular season, there are likely to be more opportunities for contact with the others that subsequently result in wounds. Creech et al. examined an American men's football team and found a high rate of SA infection during the regular season [15]. Kazakova et al. also reported an outbreak of MRSA in a men's football team during the regular season [1]. Thus, evaluations of players taking part in physical contact sports during the regular season are critical for prevention and control of SA infections. In addition, individuals with SA-infected abscesses need to be carefully evaluated, as they may also serve as an SA "reservoir" that facilitates transmission to uninfected players. Our current analyses indicated that the regular season is intrinsically associated with risk factors for SA outbreaks (**Figure 2A**). Individuals with SA-infected abscesses might also be SA reservoirs that facilitate transmission to uninfected players.

3. The role of fomites in SA infection outbreaks in athletes

Data suggest that athletes with SA skin infection are more likely to experience a recurrence if the fomites were contaminated with SA. For example, an investigation of an outbreak of a MRSA infection in two different football teams revealed that the responsible MRSA clone was not found in the nares of any of the infected players, uninfected teammates/staff, or the environment [1, 3]. In a retrospective study that examined community-onset MRSA skin infections among professional football players, Kazakova et al. [1] did not find any MRSA in the nasal swabs or environmental cultures, even though 42% of the players were nasal carriers of the MSSA strains. Apart from these highly selected populations, it remains questionable whether the results from these studies can be extrapolated to the general population [16]. These findings suggest that the strain responsible for the infection was acquired from a non-nasal endogenous source or environmental sources. Moreover, the MRSA infection observed in these outbreaks was associated with exposures to various contaminated fomites, including whirlpools, shared razors, and shared towels. Other fomites implicated in the outbreaks of sports team-associated MRSA infections include benches, body sites worn by fencers, and even a bar of soap [1, 12, 17]. In non-outbreak settings, it has been reported that close contact with a person who has a skin infection was also associated with the SA infection [11].

Aggressive control of SA strains in the environment has contributed to effective strategies that can be used to prevent SA infection. For example, the National Collegiate Athletic Association (NCAA) has implemented prevention programs [18] that encourage hand hygiene [9], surveys of environmental contamination [12], showering the entire body with an antimicrobial soap and water immediately after each practice and game [19], discouraging cosmetic body shaving [3], and cleaning and disinfecting shared items [18]. As a result of active surveillance (a way of carefully monitoring of SA nasal carriage), consensus has been reached concerning the optimal ways for controlling infection among athletes [15], with nares screening for SA critical for preventing skin and soft tissue infection.

4. Nasal carriage and SA infection in athletes

After the reports of the initial studies on SA infection, sports medicine scientists have focused on the SA nasal carriage and skin infection. In human beings, the nose is the main ecological

http://dx.doi.org/10.5772/66936

niche where SA resides [20]. The primary reservoir of SA is thought to be the anterior nares, and 30% of individuals carry nasal SA at any given time [6]. The association between SA nasal carriage and staphylococcal disease was first reported by Danbolt in 1931, who studied furunculosis [21].

Epidemiologic studies found higher SA nasal carriage rates and skin lesions in players of various sports. Sports activities that can cause skin lesions are also correlated with higher SA nasal carriage rates. These include river rafting [22] and football [1]. Decker et al. reported that higher nasal carriage rates were correlated with SA skin infections [22]. They postulated that maceration of the skin caused by prolonged contact with water in conjunction with repeated small cuts or skin injuries might have been the cause of the infections. Begier et al. [3] examined the sports activity of the players for an American football team and found that 97 of 100 players were positive for SA nasal carriage. Supporting these data is a further study that found repeated skin punctures in drug users and diabetics appeared to be the source of higher SA nasal carriage rates [6]. In addition, a retrospective analysis demonstrated that infection rates tended to peak among rugby forwards [23], American football linemen [1, 12, 14], cornerbacks, and wide receivers [3], all of whom have frequent contact with other players. All of these athletes play in the front lines, engage in frequent and aggressive skin-to-skin contact during matches, and are expected to engage the opposing team in the blind side of the scrum and tackle other players. This high-frequency, rough physical contact causes the skin abrasions that are associated with SA soft tissue infections [8].

Persistent SA nasal carriage is an established risk factor for cutaneous infection in physical contact sports [15]. Despite various proposals, there has yet to be a standard definition regarding the number of cultures that need to be taken or what fraction should be positive when determining the carrier status [24]. However, attempts to define persistent carriers have been problematic, as most SA infections originate from lines that are specific to carriers and hands, which are often the primary vectors for transmitting nasal bacteria in athletes [13]. Moreover, there are a number of infectious diseases that can be spread from one person to another by contaminated hands and body sites in athletes [13]. Therefore, the current consensus is that SA resides in the anterior nares of individuals and which serve as reservoirs that predispose players to subsequent infections.

The quantity of the SA colony-forming units (CFU) that can be recovered from swabs used to examine the noses of carriers varies widely, with numbers reported to range from the single digits to millions [25, 26]. In addition, other studies have reported that there is a strong association between high cell counts and persistent carriage [24, 27, 28]. Furthermore, evidence from various studies has led to the postulate that persistent carriers represent a separate group that is distinct from the intermittent and noncarriers [27, 29]. (Most studies that have examined SA nasal carriage have used a cross-sectional design with a single-nasal culture in order to determine whether an individual is a carrier. However, longitudinal studies have distinguished at least three SA nasal carriage patterns in healthy individuals: persistent carriage, intermittent carriage, and non-carriage [6, 30–32]. In addition, some studies make a further distinction between occasional and intermittent carriers [33, 34].) Even though the reasons remain unknown, the basic determinants of persistent and intermittent carriage are thought to be different. Persistent carriers are often colonized by a single strain of SA over long time periods, whereas intermittent carriers may carry different strains over time [31, 33, 35]. Furthermore, the load of SA is higher in persistent carriers, which results in increased dispersal and a higher risk of infection [24, 29]. Nasal carriers who are also persistent carriers are reported to have higher SA loads and disperse more SA [21, 36, 37].

Approaches that use high-throughput nasal swab data can also be applied to help in our understanding of the bacterial spread. While many studies have focused on nasal swab data, it is still unknown whether nasal SA colonization alone can trigger an SA outbreak. By achieving a deeper understanding of the repercussions of carrying nasal SA, this should help to refine and optimize strategies for risk control among athletes, thereby reducing SA infections.

5. Skin surface of *S. aureus* in athletes

The skin is the largest organ of the human body, representing more than 10% of the body mass [38]. In athletes who participate in contact and collision sports, the risk of transmission of SA has been shown to be particularly high [39–41]. It has been hypothesized that "skin-to-skin contact" might be the main cause of SA transmission in athletes, with the physical contact inducing SA dissemination in these athletes [1]. The average area of the skin surface of a human adult is 2 m^2 [42]. Although a dry, salty, low-pH skin surface discourages SA growth [43], the skin of an athlete is usually soaked in sweat, which provides a moist and nourishing environment that is suitable for SA growth. Therefore, skin sweat has been considered to be a key point of transmission during physical contact [44].

Recent evidence suggests that nasal SA has a high propensity to colonize the skin surface [45]. This idea is supported by the finding that colonization often simultaneously disappears from other body sites if an intranasal topical antibiotic is used to temporarily eliminate the SA nasal carriage [46]. Furthermore, cutaneous investigations that examined sweat glands, sebaceous glands, and hair follicles have reported that these areas are likely to be associated with their own unique microbiota [47]. Sebaceous glands secrete lipid-rich sebum, with this hydrophobic coating able to protect and lubricate the hair and skin. In general, sebum serves as an antibacterial coating and acts as a molecular defense mechanism [48]. However, the relationship between exercise-induced sweating and SA transmission in physical contact sports among athletes remains unclear.

It has been reported that the nasal cavity is the primary reservoir for SA and that these carriers are an established risk factor for transmission. Two factors may be involved in the SA transmission in an athletic setting. First, nasal carriers also carry the organism on their hands. Thus, not only are contaminated hands considered to be a likely source for causing the transmission, the hands actually serve in many cases as the primary vectors for transmitting the nasal SA. Second, SA can also live on the skin, which makes it easy to transmit from one person to another via sweat. This route is considered to be the major mode of transmission. The reason for the presence of a higher density of SA on the skin surface is due to the sweat that occurs during exercise in nasal carriers [11].

http://dx.doi.org/10.5772/66936

Even though SA is found on the skin, the nose appears to be the primary reservoir for its replication and transmission to other body sites. This hypothesis is supported by studies that have demonstrated that the use of an intranasal topical antibiotic will temporarily eliminate the transmission of the SA from the nasal carriage to the colonized body site [46]. Pulsed-field gel electrophoresis (PFGE) has also shown that nasal SA isolates are often identical to the strains that later cause clinical infections [49, 50]. Since 10% of the nasal SA carriers exhibited more than one genotype or phage type in their nose, this suggests that many of the infections might be of endogenous origin [32, 51].

Direct physical contact with bodily fluids is believed to be one source of SA dissemination [52]. Examples of direct contact in rugby occur in the scrum, when making tackles, shaking hands, or coming in contact with perspiration and skin lesions. To determine the factors behind SA transmission in physical contact sports, Suzuki and Tagami [11] examined the skin surface SA before and after exercise. The findings of this study showed that the density of the nasal SA was correlated with that of the skin surface SA in nasal carriers with perspiration on the skin surface after exercise, which indicates that perspiring during exercise promotes the appearance of SA in nasal carriers. Eda et al. also provided direct evidence of skin surface SA in healthy adult males after participating in high-intensity endurance exercises [53]. Perspiring during exercise appears to be a key part in the self-infection and transmission of SA in nasal carriers. Thus, the chances of team players transmitting SA to other team members would be increased during practices and while taking part in other exercises.

6. Genotyping

Although there is a low risk of the SA infection in team sports, early detection and an awareness of possible pathways of SA transmission could play a huge part in reducing social and economic impacts if an outbreak was to occur in a particular type of sports. Studies that have examined team sports have reported on the importance of early detection in the prevention of the spread of SA [1]. When there is an SA outbreak among a sports team, the first goal should be to identify all of the carriers, which includes both players and the coaching staff. However, it can be difficult to directly obtain such information at the present. Since the SA isolation test is the most reliable and sensitive method that can be used in these identifications, the use of these tests is essential for accurate surveillance of SA outbreaks. However, it should be noted that these tests also isolate many nonspecific SA from the anterior nares of the nose or the wound. Thus, the lack of SA specificity could hinder the surveillance. At present, both mannitol salt agar with egg yolk and Baird-Parker agar media are specifically used for SA isolation. Since these media require a large amount of time for the preparation, this raises the labor costs. In addition, the mannitol salt agar with egg yolk and Baird-Parker media exhibit weak reactivity against other different bacteria, and thus, this test requires appropriate proficiency in the discerning of the colony. Therefore, a reliable method that can be readily adopted by general diagnostic laboratories will need to be developed in order to improve the diagnostic ability of these tests. The examination methodology used is central to the SA surveillance.

Recently, molecular typing methods have greatly improved our understanding of SA transmission, provided powerful tools for tracing the transmission of individual strains and revealed methicillin-resistant SA (MRSA) strains [17]. Since there is a lack of data on the prevalence of SA transmission among athletes, this has prevented effective surveillance, thereby leading to the failure of preventing infections. In the infection control field, our understanding of SA transmission is limited by the methods used to determine the relatedness of microorganisms in the context of time and space. Conventional typing methods, such as phage typing, multi-locus sequence typing (MLST) and pulsed-field gel electrophoresis (PFGE) [54], spa typing [55, 56], and multi-locus enzyme electrophoresis (MLEE) [57], have all been successfully used to describe the global population structure of SA. In addition, this methodology has been used to provide a framework for the description of the major lineages associated with healthcare-associated infections in different countries and to monitor their emergence, dispersal, and decline in different settings [58]. However, when attempting to investigate the finer details of infection outbreaks, these conventional typing methods have serious limitations [59]. Phage-open reading frame typing (POT) has been developed as a genotyping tool based on multiplex polymerase chain reaction (PCR) [60]. These POT methods have been applied to investigate nosocomial MRSA outbreaks, with the discriminatory power of the method shown to be excellent [61, 62]. Although strategies that use molecular genotyping have been able to successfully detect the presence of SA colonization within a few hours, at the present time, they cannot help in predicting the carrier state. In addition, these methods are expensive as compared to that for standard cultures.

Acknowledgements

This paper is a revised and enhanced version of the authors' recent publication [11]. This study was funded by the Ministry of Education, Culture, Sports, Science and Technology, grant number (KAKENHI24700748), Grant-in-Aid for Young Scientists (B).

Author details

Kotaro Suzuki

Address all correspondence to: koutarousuzuki55@gmail.com

Faculty of Health and Sports Sciences, University of Tsukuba, Tsukuba, Ibaraki, Japan

References

[1] Kazakova SV, Hageman JC, Matava M, Srinivasan A, Phelan L, Garfinkel B, Boo T, McAllister S, Anderson J, Jensen B, Dodson D, Lonsway D, McDougal LK, Arduino M, Fraser VJ, Killgore G, Tenover FC, Cody S, Jernigan DB. A clone of methicillin-resistant *Staphylococcus aureus* among professional football players. N Engl J Med 2005; **352**: 468–475. DOI: 10.1056/NEJMoa042859

[2] Oller AR, Province L, Curless B. *Staphylococcus aureus* recovery from environmental and human locations in 2 collegiate athletic teams. J Athl Train 2009; **45**: 222–229. DOI: 10.4085/1062-6050-45.3.222.

[3] Begier EM, Frenette K, Barrett NL, Mshar P, Petit S, Boxrud DJ, Watkins-Colwell K, Wheeler S, Cebelinski EA, Glennen A, Nguyen D, Hadler JL. A high-morbidity outbreak of methicillin-resistant *Staphylococcus aureus* among players on a college football team, facilitated by cosmetic body shaving and turf burns. Clin Infect Dis 2004; **39**: 1446–1453. DOI: 10.1086/425313

[4] Huijsdens XW, van Lier AM, van Kregten E, Verhoef L, van Santen-Verheuvel MG, Spalburg E, Wannet WJ. Methicillin-resistant *Staphylococcus aureus* in Dutch soccer team. Emerg Infect Dis 2006; **12**: 1584–1586. DOI: 10.3201/eid1210.060387

[5] Rackham DM, Ray SM, Franks AS, Bielak KM, Pinn TM. Community-associated methicillin-resistant *Staphylococcus aureus* nasal carriage in a college student athlete population. Clin J Sport Med 2010; **20**: 185–188. DOI: 10.1097/JSM.0b013e3181dba80d000 42752-201005000-00004

[6] Kluytmans J, van Belkum A, Verbrugh H. Nasal carriage of *Staphylococcus aureus*: epidemiology, underlying mechanisms, and associated risks. Clin Microbiol Rev 1997; **10**: 505–520. http://www.ncbi.nlm.nih.gov/entrez/query.fcgi?cmd=Retrieve&db=PubMed&dopt=Citation&list_uids=9227864

[7] White A. Increased infection rates in heavy nasal carriers of coagulase-positive staphylococci. Antimicrob Agents Chemother 1963; **161**: 667–670. http://www.ncbi.nlm.nih.gov/pubmed/14274980

[8] Bartlett PC, Martin RJ, Cahill BR. Furunculosis in a high school football team. Am J Sports Med 1982; **10**: 371–374. http://www.ncbi.nlm.nih.gov/pubmed/7180958

[9] Redziniak DE, Diduch DR, Turman K, Hart J, Grindstaff TL, MacKnight JM, Mistry DJ. Methicillin-resistant *Staphylococcus aureus* (MRSA) in the athlete. Int J Sports Med 2009; **30**: 557–562. DOI: 10.1055/s-0029-1214382

[10] Sosin DM, Gunn RA, Ford WL, Skaggs JW. An outbreak of furunculosis among high school athletes. Am J Sports Med 1989; **17**: 828–832. http://www.ncbi.nlm.nih.gov/entrez/query.fcgi?cmd=Retrieve&db=PubMed&dopt=Citation&list_uids=2624295

[11] Suzuki K, Tagami K. Role of nasal *Staphylococcus aureus* carriage in transmission among contact athletes. Int J Sports Med 2015; **36**: 1186–1191. DOI: 10.1055/s-0035-1555856

[12] Nguyen DM, Mascola L, Brancoft E. Recurring methicillin-resistant *Staphylococcus aureus* infections in a football team. Emerg Infect Dis 2005; **11**: 526–532. http://www.ncbi.nlm.nih.gov/entrez/query.fcgi?cmd=Retrieve&db=PubMed&dopt=Citation&list_uids=15829189

[13] Oller AR, Province L, Curless B. *Staphylococcus aureus* recovery from environmental and human locations in 2 collegiate athletic teams. J Athl Train 2010; **45**: 222–229. DOI: 10.4085/1062-6050-45.3.222.

[14] Romano R, Lu D, Holtom P. Outbreak of community-acquired methicillin-resistant *Staphylococcus aureus* skin infections among a collegiate football team. J Athl Train 2006; **41**: 141–145. http://www.ncbi.nlm.nih.gov/entrez/query.fcgi?cmd=Retrieve&db=PubMed&dopt=Citation&list_uids=16791297

[15] Creech CB, Saye E, McKenna BD, Johnson BG, Jimenez N, Talbot TR, Bossung T, Gregory A, Edwards KM. One-year surveillance of methicillin-resistant *Staphylococcus aureus* nasal colonization and skin and soft tissue infections in collegiate athletes. Arch Pediatr Adolesc Med 2010; **164**: 615–620. DOI: 10.1001/archpediatrics.2010.93

[16] Ellis MW, Hospenthal DR, Dooley DP, Gray PJ, Murray CK. Natural history of community-acquired methicillin-resistant *Staphylococcus aureus* colonization and infection in soldiers. Clin Infect Dis 2004; **39**: 971–979. DOI: 10.1086/423965

[17] CDC. Methicillin-resistant *Staphylococcus aureus* infections among competitive sports participants – Colorado, Indiana, Pennsylvania, and Los Angeles County, 2000–2003. MMWR Morb Mortal Wkly Rep 2003; **52**: 793–795.

[18] Zinder SM, Basler RS, Foley J, Scarlata C, Vasily DB. National athletic trainers' association position statement: skin diseases. J Athl Train 2010; **45**: 411–428. DOI: 10.4085/1062-6050-45.4.411

[19] Cordoro KM, Ganz JE. Training room management of medical conditions: sports dermatology. Clin Sports Med 2005; **24**: 565–598. DOI: 10.1016/j.csm.2005.03.008

[20] Peacock SJ, de Silva I, Lowy FD. What determines nasal carriage of *Staphylococcus aureus*? Trends Microbiol 2001; **9**: 605–610. DOI: 10.1016/S0966-842X(01)02254-5

[21] Solberg CO. A study of carriers of *Staphylococcus aureus* with special regard to quantitative bacterial estimations. Acta Med Scand Suppl 1965; **436**: 1–96. http://www.ncbi.nlm.nih.gov/pubmed/5318930

[22] Decker MD, Lybarger JA, Vaughn WK, Hutcheson RH, Jr., Schaffner W. An outbreak of staphylococcal skin infections among river rafting guides. Am J Epidemiol 1986; **124**: 969–976. http://www.ncbi.nlm.nih.gov/pubmed/3776979

[23] Stacey AR, Endersby KE, Chan PC, Marples RR. An outbreak of methicillin resistant *Staphylococcus aureus* infection in a rugby football team. Br J Sports Med 1998; **32**: 153–154. http://www.ncbi.nlm.nih.gov/entrez/query.fcgi?cmd=Retrieve&db=PubMed&dopt=Citation&list_uids=9631224

[24] Nouwen JL, Ott A, Kluytmans-Vandenbergh MF, Boelens HA, Hofman A, van Belkum A, Verbrugh HA. Predicting the *Staphylococcus aureus* nasal carrier state: derivation and validation of a "culture rule". Clin Infect Dis 2004; **39**: 806–811. DOI: 10.1086/423376

[25] Mermel LA, Eells SJ, Acharya MK, Cartony JM, Dacus D, Fadem S, Gay EA, Gordon S, Lonks JR, Perl TM, McDougal LK, McGowan JE, Maxey G, Morse D, Tenover FC. Quantitative analysis and molecular fingerprinting of methicillin-resistant *Staphylococcus*

aureus nasal colonization in different patient populations: a prospective, multicenter study. Infect Control Hosp Epidemiol 2010; **31**: 592–597. DOI: 10.1086/652778

[26] White A. Quantitative studies of nasal carriers of staphylococci among hospitalized patients. J Clin Invest 1961; **40**: 23–30. DOI: 10.1172/JCI104233

[27] van Belkum A, Verkaik NJ, de Vogel CP, Boelens HA, Verveer J, Nouwen JL, Verbrugh HA, Wertheim HF. Reclassification of *Staphylococcus aureus* nasal carriage types. J Infect Dis 2009; **199**: 1820–1826. DOI: 10.1086/599119

[28] Verhoeven PO, Grattard F, Carricajo A, Lucht F, Cazorla C, Garraud O, Pozzetto B, Berthelot P. Quantification by real-time PCR assay of *Staphylococcus aureus* load: a useful tool for rapidly identifying persistent nasal carriers. J Clin Microbiol 2012; **50**: 2063–2065. DOI: 10.1128/JCM.00157-12

[29] Nouwen JL, Fieren MW, Snijders S, Verbrugh HA, van Belkum A. Persistent (not intermittent) nasal carriage of *Staphylococcus aureus* is the determinant of CPD-related infections. Kidney Int 2005; **67**: 1084–1092. DOI: 10.1111/j.1523-1755.2005.00174.x

[30] Armstrong-Esther CA. Carriage patterns of *Staphylococcus aureus* in a healthy non-hospital population of adults and children. Ann Hum Biol 1976; **3**: 221–227. http://www.ncbi.nlm.nih.gov/pubmed/962302

[31] VandenBergh MF, Yzerman EP, van Belkum A, Boelens HA, Sijmons M, Verbrugh HA. Follow-up of *Staphylococcus aureus* nasal carriage after 8 years: redefining the persistent carrier state. J Clin Microbiol 1999; **37**: 3133–3140. http://www.ncbi.nlm.nih.gov/pubmed/10488166

[32] Williams RE. Healthy carriage of *Staphylococcus aureus*: its prevalence and importance. Bacteriol Rev 1963; **27**: 56–71. http://www.ncbi.nlm.nih.gov/pubmed/14000926

[33] Eriksen NH, Espersen F, Rosdahl VT, Jensen K. Carriage of *Staphylococcus aureus* among 104 healthy persons during a 19-month period. Epidemiol Infect 1995; **115**: 51–60. http://www.ncbi.nlm.nih.gov/pubmed/7641838

[34] Maxwell JG, Ford CR, Peterson DE, Mitchell CR. Long-term study of nasal staphylococci among hospital personnel. Am J Surg 1969; **118**: 849–854. http://www.ncbi.nlm.nih.gov/pubmed/4243232

[35] Hu L, Umeda A, Kondo S, Amako K. Typing of *Staphylococcus aureus* colonising human nasal carriers by pulsed-field gel electrophoresis. J Med Microbiol 1995; **42**: 127–132. DOI: 10.1099/00222615-42-2-127

[36] Ridley M. Perineal carriage of *Staph. aureus*. Br Med J 1959; **1**: 270–273. http://www.ncbi.nlm.nih.gov/pubmed/13618615

[37] Squier C, Rihs JD, Risa KJ, Sagnimeni A, Wagener MM, Stout J, Muder RR, Singh N. *Staphylococcus aureus* rectal carriage and its association with infections in patients in a surgical intensive care unit and a liver transplant unit. Infect Control Hosp Epidemiol 2002; **23**: 495–501. DOI: 10.1086/502095

[38] Crosera M, Bovenzi M, Maina G, Adami G, Zanette C, Florio C, Filon Larese F. Nanoparticle dermal absorption and toxicity: a review of the literature. Int Arch Occup Environ Health 2009; **82**: 1043–1055. DOI: 10.1007/s00420-009-0458-x

[39] Braun T, Kahanov L, Dannelly K, Lauber C. CA-MRSA infection incidence and care in high school and intercollegiate athletics. Med Sci Sports Exerc 2016; **48**: 1530–1538. DOI: 10.1249/MSS.0000000000000940

[40] Collins CJ, O'Connell B. Infectious disease outbreaks in competitive sports, 2005–2010. J Athl Train 2012; **47**: 516–518. DOI: 10.4085/1062-6050-47.5.02

[41] Jimenez-Truque N, Saye EJ, Soper N, Saville BR, Thomsen I, Edwards KM, Creech CB. Longitudinal assessment of colonization with *Staphylococcus aureus* in healthy collegiate athletes. J Pediatric Infect Dis Soc 2016; **5**: 105–113. DOI: 10.1093/jpids/piu108

[42] Tobin DJ. Biochemistry of human skin—our brain on the outside. Chem Soc Rev 2006; **35**: 52–67. DOI: 10.1039/b505793k

[43] Korting HC, Lukacs A, Vogt N, Urban J, Ehret W, Ruckdeschel G. Influence of the pH-value on the growth of *Staphylococcus epidermidis*, *Staphylococcus aureus* and *Propionibacterium acnes* in continuous culture. Int J Hyg Environ Med 1992; **193**: 78–90. http://www.ncbi.nlm.nih.gov/pubmed/1503605

[44] Weber K. Community-Associated Methicillin-Resistant *Staphylococcus aureus* Infections in the Athlete. Sports Health. 2009; **1**: 405-410. DOI: 10.1177/1941738109343653

[45] Wertheim HF, Melles DC, Vos MC, van Leeuwen W, van Belkum A, Verbrugh HA, Nouwen JL. The role of nasal carriage in *Staphylococcus aureus* infections. Lancet Infect Dis 2005; **5**: 751–762. DOI: 10.1016/S1473-3099(05)70295-4

[46] Reagan DR, Doebbeling BN, Pfaller MA, Sheetz CT, Houston AK, Hollis RJ, Wenzel RP. Elimination of coincident *Staphylococcus aureus* nasal and hand carriage with intranasal application of mupirocin calcium ointment. Ann Intern Med 1991; **114**: 101–106. http://www.ncbi.nlm.nih.gov/pubmed/1898585

[47] Kong HH, Segre JA. Skin microbiome: looking back to move forward. J Invest Dermatol 2012; **132**: 933–939. DOI: 10.1038/jid.2011.417

[48] Belkaid Y, Tamoutounour S. The influence of skin microorganisms on cutaneous immunity. Nat Rev Immunol 2016; **16**: 353–366. DOI: 10.1038/nri.2016.48

[49] von Eiff C, Becker K, Machka K, Stammer H, Peters G. Nasal carriage as a source of *Staphylococcus aureus* bacteremia. Study group. N Engl J Med 2001; **344**: 11–16. DOI: 10.1056/NEJM200101043440102

[50] Wertheim HF, Vos MC, Ott A, van Belkum A, Voss A, Kluytmans JA, van Keulen PH, Vandenbroucke-Grauls CM, Meester MH, Verbrugh HA. Risk and outcome of nosocomial *Staphylococcus aureus* bacteraemia in nasal carriers versus non-carriers. Lancet 2004; **364**: 703–705. DOI: 10.1016/S0140-6736(04)16897-9

[51] Cespedes C, Said-Salim B, Miller M, Lo SH, Kreiswirth BN, Gordon RJ, Vavagiakis P, Klein RS, Lowy FD. The clonality of *Staphylococcus aureus* nasal carriage. J Infect Dis 2005; **191**: 444–452. DOI: 10.1086/427240

[52] Luke A, d'Hemecourt P. Prevention of infectious diseases in athletes. Clin Sports Med 2007; **26**: 321–344. DOI: 10.1016/j.csm.2007.04.006

[53] Eda N, Shimizu K, Suzuki S, Lee E, Akama T. Effects of high-intensity endurance exercise on epidermal barriers against microbial invasion. J Sports Sci Med 2013; **12**: 44–51. http://www.ncbi.nlm.nih.gov/pubmed/24149724

[54] Lin J, Lin D, Xu P, Zhang T, Ou Q, Bai C, Yao Z. Non-hospital environment contamination with *Staphylococcus aureus* and methicillin-resistant *Staphylococcus aureus*: proportion meta-analysis and features of antibiotic resistance and molecular genetics. Environ Res 2016; **150**: 528–540. DOI: 10.1016/j.envres.2016.06.040

[55] Aguadero V, Gonzalez Velasco C, Vindel A, Gonzalez Velasco M, Moreno JJ. Evaluation of rep-PCR/DiversiLab versus PFGE and spa typing in genotyping methicillin-resistant *Staphylococcus aureus* (MRSA). Br J Biomed Sci 2015; **72**: 120–127. http://www.ncbi.nlm.nih.gov/pubmed/26510268

[56] Katakweba AS, Muhairwa AP, Espinosa-Gongora C, Guardabassi L, Mtambo MM, Olsen JE. Spa typing and antimicrobial resistance of *Staphylococcus aureus* from healthy humans, pigs and dogs in Tanzania. J Infect Dev Ctries 2016; **10**: 143–148. DOI: 10.3855/jidc.6790

[57] Fitzgerald JR, Meaney WJ, Hartigan PJ, Smyth CJ, Kapur V. Fine-structure molecular epidemiological analysis of *Staphylococcus aureus* recovered from cows. Epidemiol Infect 1997; **119**: 261–269. http://www.ncbi.nlm.nih.gov/pubmed/9363026

[58] Price JR, Didelot X, Crook DW, Llewelyn MJ, Paul J. Whole genome sequencing in the prevention and control of *Staphylococcus aureus* infection. J Hosp Infect 2013; **83**: 14–21. DOI: 10.1016/j.jhin.2012.10.003

[59] Harris SR, Clarke IN, Seth-Smith HM, Solomon AW, Cutcliffe LT, Marsh P, Skilton RJ, Holland MJ, Mabey D, Peeling RW, Lewis DA, Spratt BG, Unemo M, Persson K, Bjartling C, Brunham R, de Vries HJ, Morre SA, Speksnijder A, Bebear CM, Clerc M, de Barbeyrac B, Parkhill J, Thomson NR. Whole-genome analysis of diverse *Chlamydia trachomatis* strains identifies phylogenetic relationships masked by current clinical typing. Nat Genet 2012; **44**: 413–419, S411. DOI: 10.1038/ng.2214

[60] Suzuki M, Tawada Y, Kato M, Hori H, Mamiya N, Hayashi Y, Nakano M, Fukushima R, Katai A, Tanaka T, Hata M, Matsumoto M, Takahashi M, Sakae K. Development of a rapid strain differentiation method for methicillin-resistant *Staphylococcus aureus* isolated in Japan by detecting phage-derived open-reading frames. J Appl Microbiol 2006; **101**: 938–947. DOI: 10.1111/j.1365-2672.2006.02932.x

[61] Nada T, Yagi T, Ohkura T, Morishita Y, Baba H, Ohta M, Suzuki M. Usefulness of phage open-reading frame typing method in an epidemiological study of an outbreak of methicillin-resistant *Staphylococcus aureus* infections. Jpn J Infect Dis 2009; **62**: 386–389. http://www.ncbi.nlm.nih.gov/pubmed/19762991

[62] O'Sullivan MV, Kong F, Sintchenko V, Gilbert GL. Rapid identification of methicillin-resistant *Staphylococcus aureus* transmission in hospitals by use of phage-derived open reading frame typing enhanced by multiplex PCR and reverse line blot assay. J Clin Microbiol 2010; **48**: 2741–2748. DOI: 10.1128/JCM.02201-09

Chapter 3

Methicillin-Resistant *Staphylococcus aureus* (MRSA) in Food-Producing and Companion Animals and Food Products

Jungwhan Chon, Kidon Sung and Saeed Khan

Additional information is available at the end of the chapter

http://dx.doi.org/10.5772/66645

Abstract

Methicillin-resistant *Staphylococcus aureus* (MRSA) has become a growing concern in companion and food-producing animals. The presence of multidrug-resistance with a wide range of extracellular enterotoxin genes, virulence factors, and Panton-Valentine leukocidin (*pvl*) cytotoxin genes confer life-threatening traits on MRSA and makes them highly pathogenic and difficult to treat. Clonal complex 398 (CC398), a predominant clonal lineage of livestock-associated-MRSA in domestic animals and retail meat, is capable of infecting humans. In order to monitor and prevent MRSA contamination, it is critical to understand its source and transmission dynamics. In this review, we describe MRSA in food-producing animals (pig, cattle, chicken), horses, pet animals (dogs, cats), and food products (pork, beef, chicken, milk, and fish).

Keywords: MRSA, companion animals, food-producing animals, food products, CC398

1. Introduction

Staphylococcus aureus is a pathogen that causes both human and animal infections and food intoxication [1–3]. It causes simple infections, such as furuncle, boil, stye, impetigo, carbuncle, and keratitis, and serious infections, including septicemia, necrotizing pneumonia, endocarditis, osteomyelitis, and pericarditis [4–7]. Shortly after the introduction of methicillin in clinical practice to control penicillin-resistant staphylococci, the first methicillin-resistant *S. aureus* (MRSA) was isolated [8]. MRSA is one of the most important hospital-acquired pathogens that are resistant to various antimicrobials, thereby making their treatment complicated [9]. MRSA in humans is usually divided into two groups: hospital-associated MRSA (HA-MRSA) and community-associated MRSA (CA-MRSA) [10]. A third group of MRSA, known as livestock-associated MRSA (LA-MRSA), now has emerged and infects livestock, pets, and wild animals.

LA-MRSA was first detected in milk with bovine mastitis from Belgium in 1972 [11–13]. Thereafter, MRSA reports in various food and companion animals, such as pigs, cattle, chickens, dogs, cats, and horses, have increased [11, 14]. A novel strain of MRSA belonging to multi-locus sequencing type (MLST) 398 (ST398) and related strains collectively grouped into clonal complex 398 (CC 398) have been frequently found in pigs, chickens, veal calves, dairy cattle, horses, dogs, and milk in various countries [11]. Both methicillin-susceptible *S. aureus* (MSSA) and MRSA have been associated with companion and food production animals [15–19]. The most significant of these is intramammary infection of dairy cattle leading to mastitis, which causes a substantial economic loss to the dairy industry worldwide [6, 20, 21]. The CC398 *S. aureus* isolate was more prevalent in nasal swabs of pig and cattle farmers than of non-farming human controls [22, 23]. An examination of livestock-associated MRSA (LA-MRSA) in human case isolates in the Netherlands indicated an increase from 0% in 2002 to greater than 21% in mid-2006 [23] and 35% in 2009 [24]. In most European countries, CC398 remains the most commonly identified type of LA-MRSA [15, 25–27]. However, the epidemiology of LA-MRSA differs in other geographic areas. A different strain of LA-MRSA, CC9, appears to be the prominent type in several Asian countries [28–32]. Poultry may harbor CC398 strains [16, 33, 34] but CC5 [33, 35] and other types unrelated to CC398 have also been reported [36]. The diversity of LA-MRSA in the USA appears to be higher than that identified in Europe or Asia, with reports of both CC398 as well as a variety of "human" types of *S. aureus* in livestock.

LA-MRSA infections among livestock animals and associated farmers are of great concern as these sources could potentially serve as reservoirs for zoonotic infections [14]. Contamination of food with enterotoxin producing *S. aureus* leads to over 240,000 cases of food-borne illness in the United States annually. Although most *S. aureus*–related food-poisoning incidents are self-limiting and go away within 2 days, some serious infections have been reported as well [4, 5]. A large number of the reported staphylococcal food-poisoning outbreaks can be traced back to a human source harboring *S. aureus* producing certain staphylococcal enterotoxins (SEs) [1, 37]. Most of the LA-MRSA strains, particularly the ST398 group, do not appear to code for any of the known SEs [11, 38–42]. However, genes for SEs B, K, and Q have been detected in MRSA CC398 strains isolated from geographically diverse pig farms in Germany [43]. The acquisition of enterotoxin genes along with the virulence factors, such as Panton-Valentine leukocidin (*pvl*) genes by LA-MRSA may eventually pose a threat to humans, suggesting that animals have the potential to be a source of primary contamination as well [16, 44].

2. MRSA in food-producing animals

2.1. Porcine MRSA

In 2005, pigs were reported as an animal reservoir of *S. aureus* in France, including MRSA [17]. Pig farmers were more frequently colonized than nonfarmers and one of the most prevalent strains of *S. aureus* in pig farmers was CC398 [45–49]. Forty-five percent of veterinarians attending pig farms in the Netherlands were positive for MRSA [48]. In Belgium and Denmark, the prevalence of MRSA in veterinarians was 9.5% and 1.4%, respectively [47]. German researchers reported that MRSA ST398 that carried SCC*mec* IV or V, accessory gene regulator type I and capsule type 5 [25] in pig primary production herds was higher in their country (45–70%) than in the rest of the European Union [45, 46]. In an interesting report, 12.5% of attendees at an international meeting concerning pig health carried MRSA and 91.2% (31/34) of them CC398 [50]. While early studies on farms and of meat identified CC398 strains in animals, farm workers, and meat products [51, 52], other studies also documented CC398 in populations with no obvious livestock contact [53–55]. The emergence of this strain was also reported in pigs and pig farmers in the Netherlands [22, 56, 57], Denmark [58], Germany [59], and Canada [60]. Detection rates of MRSA in breeding and production herds were 46%, 43.5%, and 40% in Spain, Germany, and Belgium, respectively, but no single strain of MRSA was found in Finland and Denmark in 2008 [61]. The majority of LA-MRSA lineage belonged to ST398, accounting for 92.5% of the MRSA isolates. Other ST types, human-associated MRSA ST1 and cattle-associated ST97 in finishing holdings, and ST9 in the same animal species in Europe [62], CC9, and CC49 in Switzerland [63] have also been reported.

Various farm types in the Netherlands were reported to have MRSA in 23–71% of their pigs, and it was especially high in farms with finishing pigs (pigs are almost ready to be sent to market) [49, 57, 64]. The presence of MRSA is dependent on pig production type and herd size and increases from 31 to 86% depending upon small-, medium-, and large-sized farms carrying <250, >500, and >1000 animals, respectively [45, 62, 64, 65]. MRSA prevalence also varied with farm type, e.g., fattening and closed (farrow-to-finish) farms exhibited 94 and 56% MRSA, respectively [66]. Transportation from farm to slaughterhouse [38, 67], lairage [38], national and international trade [57], and slaughter house employees all have been reported to enhance MRSA contamination and may play important roles in transmission of the bacteria [68]. It was proposed that MRSA contamination in piglets is dependent on the status of sows [69]. When a sow was colonized with MRSA, 100% piglets were MRSA-positive. However, 84% piglets were MRSA-positive when there was no MRSA contamination in a sow. Higher numbers of MRSA were isolated in suckling (52.9%) and weanling piglets (53.4%) than sows (38.3%) [64]. Prevalence of MRSA in pigs has been linked to their age but the data are not conclusive. MRSA were identified in 100% of 9–12-week-old pigs, whereas in adult animals it decreased to 36% [51]. On the other hand, Weese et al. [69] reported that MRSA was more prevalent in post-weaning (85%) than preweaning pigs (34.5%) in pig farms without antimicrobial treatment. Percent of MRSA colonization in Canadian piglets on days 1, 28, 56, and 70 were 1, 34, 50, and 42%, respectively [69]. Khanna et al. [60] reported no variation in MRSA prevalence based on age groups. MRSA does not seem to cause serious infection in pigs, but there have been a few reports of MRSA from exudative epidermitis lesions of piglets on a breeding farm [70] and in pigs suffering from infection of the urinary-genital tract, skin infection, and metritis-mastitis-agalactia syndrome [71]. While MRSA ST398 isolated from diseased pigs did not carry the major virulence genes, such as toxic shock syndrome toxin 1, *pvl*, and exfoliative toxins, they carried some virulence genes, such as α- and δ-hemolysins, proteases, capsule type-specific genes, microbial surface components recognizing adhesive matrix molecules, biofilm-associated, and enterotoxin genes [44]. Their MLST, *spa*, and SCC*mec* types were identified as ST398, t011, and IV, respectively.

The epidemiology of livestock-associated *S. aureus* in the USA appears to be notably different than that in European countries. Most of the porcine MRSA isolates in Canada, Europe, Peru, and USA were CC398 [65]. A human epidemic clone, Canadian MRSA-2 (CMRSA-2, USA100, CC5), was found in nasal and rectal swabs of pigs in Canada [60]. This isolate was the most common cause of health care-associated infections in Canada. CMRSA-5 (USA500, ST8) also isolated from retail pork in Canada, is a human epidemic strain that also has been documented in horses and horse personnel [72]. Three *spa* types (t011, t034, t108) within CC398 were the most frequent in breeding and production herds in Europe, and t108 was most popular only in the Netherlands among European countries. On the contrary, prevalence of t899 in Italian breeding and production herds was between 24% and 27% [62, 65]. In Italy, CC1 and CC97 lineages among MLST types that do not belong to CC398 were significantly high in the primary production of pigs [62]. In one study carried out in a jail setting in Texas, CC398 isolates made up of 13.2% of all MSSA identified within this population. Apart from CC398, other human strains of *S. aureus* have also been found in USA livestock. Studies carried out on swine farms in the USA have identified human strains within the noses of live animals [55, 73, 74] and farm dust [75]. Several papers have reported CC5 strains rather than CC398-associated types to be the dominant strain isolated from pig farms in both Iowa and Ohio [75, 76], whereas others have found CC398 to be the most common molecular type [51, 76]. Studies on workers on pig farms and in processing plants found substantial diversity within *S. aureus* isolates, including CC398, CC5, and CC8 strains, among others [77–79]. MRSA attributed to ST5 was recently reported in pigs in the USA [38]. A different swine-associated MRSA strain, CC9, is circulating among pigs and pig farmers in China and Malaysia [29, 30, 80, 81]. MRSA ST22, known as human epidemic clone EMRSA-15 in the UK, was also discovered in pigs and in hospitalized patients in Singapore with an elevated frequency [82, 83]. High frequency of MRSA CC9 with *spa* types t899 and t4358 was reported in porcine samples in Asian countries, whereas it was not found as much in Europe or the USA [28–30, 84]. In Thailand, porcine MRSA ST9 isolates had a unique *spa* type (t337) and SCC*mec* type (SCC*mec* IX) that were different from other LA-MRSA ST9 strains found in Asian countries [28]. In 2012, Lim et al. [85] reported two ST types, livestock-associated ST398 and human-associated ST72, from pigs, which was the first finding of ST398 in Korean pigs.

2.2. Bovine MRSA

The first report of MRSA in farm animals was published in the early 1970s, when the bacteria were isolated from the milk of dairy cows with mastitis in Belgium [86] and clustered in the CC398 group [87]. Devriese and Hommez [88] suspected that these samples were most likely contaminated by humans. In the past few years, MRSA has been isolated from cows or their milk in Korea [89–91], Hungary, Mexico, and the Netherlands [89, 92, 93]. There have also been numerous reports of MRSA from cows or their milk in Brazil, Italy, Pakistan, Nigeria, Turkey, and the USA [94–96]. Subsequently, several reports have described bovine udder infections caused by LA-MRSA CC398 [97]. In Dutch farms, MRSA was detected in 18–31% of veal calves [98]. In 2010, the European Union reported that 20% of veal calves in Germany carried MRSA [99]. A survey of 51 veal calf farms in the Netherlands indicated that an average of 38% of farmers and 16% of family members were colonized with LA-MRSA [19, 100]. Recently, another group of LA-MRSA strains (CC130, CC425, and CC1943) that were initially thought to be bovine-specific lineages emerged in humans [101]. A number of multidrug-resistant MRSA were isolated from bovine mastitis in Germany [97, 102], the majority of which were MRSA ST398 related to animal strains, but an isolate of the clonal complex group CC8 was identified as a human epidemic MRSA strain Irish-01 [103]. In 2010, Hata et al. first discovered MRSA in cow's milk in Japan and the genotypes (ST5-SCC*mec* II) were the same as or similar to human strains [104]. There has been a dramatic surge in human CC398 infection and colonization in the Netherlands, increasing from 0% in 2002 to more than 21% in 2006 [23]. A recent Dutch study indicated that the annual incidence of MRSA in humans more than tripled from 2001 to 2006 where 23%, 26%, 16%, and 10% of the patients acquired MRSA from a foreign hospital, animals, nosocomial transmission, and the community, respectively [105]. The presence of MRSA CC398 in pig farms with a concomitant increase in CC398 infections in humans clearly suggests that pigs or cattle are specifically a risk factor and CC398 MRSA colonization and prevalence in humans is associated with animal contact [23, 46, 106, 107]. The above argument is further strengthened by the findings that the CC398 carrier status of farm workers decreased dramatically when they took a break from direct animal care duties [19, 100]. People who visited farms to collect samples for a shorter duration carried MRSA transiently as compared to those who had prolonged visits, suggesting that a prolonged contact with animals is probably an important factor for higher rates of colonization [49].

While a majority of the MRSA collected from dairy cattle belonged to ST398 [89], other ST types, such as ST1-t286-SCC*mec* IV, ST72-t324 [108], ST59-t437–V [91], ST10-t127-SCC*mec* IVa genotype [92], SCC*mec* types IVg [109], CC97, t4795, and t1730 [110], and a *mecA* variant ($mecA_{LGA251}$) known as *mecC*, are also reported from MRSA CC130 and ST425 isolates [111]. The *mecC* type was also detected in Danish MRSA CC130 isolated from a cow and the genotypic characteristics, such as *spa* type (t843), MLVA (MT429) and PFGE profiles of bovine isolates were the same as the human isolates, implying transmission between humans and ruminants [112]. The geographic variation in the prevalence and origin of CC398 colonization and incidence of infection in humans is quite interesting. MRSA has been detected in retail beef, but nasal and fecal sampling of nearly 500 Canadian feedlot cattle, shortly before slaughter, detected no MRSA [113]. Moreover, while CC398 MRSA infection in humans is a leading cause of CA-MRSA infection in some European countries, it is rare in North America despite the presence of CC398 in livestock [114]. The reasons for the low incidence of CC398 infections in the USA may include differences in direct and indirect contact with food animals, much lower population density in North American pig-rearing regions, and the common presence of other competing MRSA strains in people in the general population. Although some studies suggested that the MRSA present in cattle is bovine-specific, most of the reports indicated that MRSA found in cattle were derived from humans [89, 92, 115, 116]. Bovine, porcine, canine, feline, and equine MRSA isolates containing the *pvl* gene and other virulence factors, such as *chp, scn, seb, sek,* and *seq,* toxic shock syndrome toxin 1 (*tsst1* or *tst*) gene [91, 117], hemolysin, protease, superantigen-like protein, capsule, and biofilm-associated genes [118–121], may pose a potential threat to public health.

2.3. Poultry MRSA

CC398 is not limited to large livestock animals alone; it has also been reported in poultry [18, 122], manure from chicken farms and soil fertilized with this manure [123]. However, the numbers of MRSA ST398-t011-SCC*mec* V isolated from chickens are lower (0–28%) than in pigs (82–92%) on the same farm [124]. In another study from Belgium [109], MRSA present in 0.8–1.8% layers and broilers (chickens raised for meat) was clustered into two ST types, ST398 (t011, t899) and ST239 (t037). Two other studies from Belgian broiler farms [18, 125] reported 12.8–14.3% of randomly selected Belgian broiler farms to be positive for CC398. In 2010, the Federal Institute for Risk Assessment reported a contamination rate of 32% in turkey meat in Berlin, Germany [126]. Similar contamination frequencies were reported from Canada and the USA [72, 127], as well as from Taiwan [128]. Moon et al. [91] reported that the *S. aureus* isolated from chicken carcasses contained 1.3% MRSA, which was more than the MRSA isolated from other animal carcasses (0.3%) in Korea. Poultry-associated *S. aureus* isolates belonging to genotypes other than CC398 have also been reported from different geographic regions [36, 129–131]. Mulders et al. [132] reported that 6.9% of MRSA present in broiler chickens in the Netherlands represented the ST9-t1430 genotype. A single *spa* type, t1456, in poultry was seen and distinguishable from the *spa* types of ST398 observed in other animals in Belgium [125]. Genotypic and antimicrobial patterns between 14 MRSA isolates from broilers and pigs were identical [133]. The MRSA isolates from Korea had the genotype of ST692-t2247–III [91] and the MRSA isolates from Hong Kong exhibited the genotypes ST9-t899–IV [81] and CC9 (t899, t1234) [134]. A study from Denmark analyzed the isolates from infected poultry and detected a predominant common human epidemic clone CC5 [129]. Using a population genomics approach, Lowder et al. [36] examined the origin of *S. aureus* isolates from diseased and healthy poultry from four continents and found that the majority of isolates belonged to a single clonal complex CC5 belonging to a known human-associated lineage. The poultry isolates were more closely related to each other than to human CC5 isolates, but were most similar to a subclade of CC5 that was circulating in Polish hospitals in the 1980s. In a study conducted in Korea, 930 food samples were collected, and four strains of the CA-MRSA CC5 human clone were identified [135]. In a human case study, a 63-year-old Dutch woman who owned a chicken farm developed a life-threatening endocarditis; the infecting MRSA isolate was identified as CC398 [136], similar to an isolate found in a pig farm nearby and to MRSA isolates previously found in other pig farms in the Netherlands.

2.4. Other meat products

S. aureus is found frequently in a variety of retail meat products. A Dutch Food Safety Agency analyzed 2217 samples of various kinds of meats from the retail stores and found that 11.9% of 2217 samples had MRSA [137]. The distribution of MRSA within various meat types was listed as follows: beef, 10.6%; veal, 15.2%; lamb and mutton, 6.2%; pork, 10.7%; chicken, 16.0%; turkey, 35.3%; fowl, 3.4%; and game, 2.2%. Of all the MRSA isolates, 85% of the isolates belonged to ST398; the other STs were possibly of human origin [137]. Another Dutch survey found that 46% of retail meat samples contained *S. aureus* strains, of which two (2%) were MRSA: one was CC398 and the other was USA300 [138]. Studies in Switzerland and Japan showed the prevalence of *S. aureus* in meat products to be 23 and 65%, respectively [122, 139].

A USA survey of 120 retail meat samples indicated that 39.2% contained *S. aureus* strains, 5% of which were MRSA of the types USA100 (ST5) and USA300 (ST8) [127]. A Canadian survey [72] found that 7.7% of retail meat samples harbored MRSA; 30% belonged to the clonal complex CC398, 40% were CC8, and 30% were CC5, a strain commonly found in humans in both the USA and Canada.

2.5. Milk

In general, the occurrence of MRSA in bovine mastitis isolates is well studied and its prevalence seems to be very low [140]. Following the initial reports of isolation of MRSA from mastitic cows [86], sporadic cases of MRSA in dairy cattle were detected among *S. aureus* isolates from clinical or subclinical mastitis. In one of the studies from Korea [109], MRSA were isolated from the milk of cows with an isolation ratio of 0.18%. In one report on dairy farms in Belgium, a high percentage (15%) of MRSA was found in lactating cows [140]; these cows had a previous history of MRSA. The long-term low prevalence of MRSA mastitis is quite surprising, given the number of years since the first identification of MRSA in cattle and the close contact of humans with the udders of dairy cattle. In Germany, the highest proportion of positive samples (45%) was found in nasal swabs from veal calves at slaughter and the lowest rate was 4.1% in bulk tank milk. Most isolates, irrespective of origin, were from *spa* types t011 and t034 belonging to the clonal complex CC398 [141]. The finding of LA-MRSA CC398 in tank milk suggests udder colonization and possibly cases of subclinical mastitis in dairy cattle in Germany [141]. Close contact of dairy cattle with humans could lead to a transfer of strains between them. In one of the reports from Hungary, MRSA isolates from mastitic cows and a worker were found identical by phenotypic and genotypic analysis indicating a transfer between cows and human [92].

2.6. Fish

Fish is not a normal host for staphylococci and its presence on fish is either due to disease in the fish, contamination, or poor personnel hygiene. The first report of the isolation of MRSA from Tilapia was published in 2010, where 559 *S. aureus* isolates from the brain, eyes, and kidneys of tilapia from 11 farms collected for a period of 2 years were analyzed and 50% were identified as MRSA [142]. In another study [143] from Korea that analyzed 165 *S. aureus* strains isolated from different food samples between 2003 and 2006, four were identified as MRSA. Two of these were from beef and two from fish. The two fish isolates, one from sea bass and other from rockfish, were identified as ST1 and ST72, respectively. An analysis of 200 ready to eat (RTE) fish samples collected from 10 shops belonging to four supermarket chains in Japan, 5 were MRSA and 5 others were identified as coagulase-negative MRSA [144]. Molecular typing of two MRSA isolates by *spa* sequencing and MLST typing identified t1767 and ST8, respectively. Interestingly, MRSA ST8 strains have been predominantly isolated from humans in the USA and Europe but are of rare occurrence in foods in Japan. It is not certain if the MRSA in fish was from human or fish origin. In another report from Greece, one hundred samples from RTE fish products were examined and two were reported to have MRSA belonging to the *spa* types t316 (ST359) and t548 (ST5) [145]. In a recent report, a patient developed foot infection with MRSA after a fish pedicure [146], but the origin of the MRSA could not be determined in this case. Reports of fish from Egypt, India, and Yemen have also been reported to harbor as much as 3.5% MRSA [147]. In this report, two of the MRSA isolates were found to harbor the enterotoxin genes *seg* and *sei*. Since the global spread of multi-drug-resistant bacteria has increased in the past decade, the finding of enterotoxigenic MRSA in fish should be of concern. The global trade of fish increases the possibility of intercontinental transmission of multidrug-resistant and enterotoxigenic *S. aureus* and its potential influence on consumer health worldwide should be monitored.

3. MRSA in food processing environment

After carcasses leave the slaughter-house chillers, residual MRSA on carcass surfaces can be transmitted during further processing through human hands, cutting tools, and any surfaces with direct meat contact. Manual handling during processing also can facilitate the entry of human MRSA strains into the production units. Recent surveillance data suggest that 22.5–64.8% of retail beef, pork, chicken, and turkey meats in five different geographical locations in the United States were contaminated with *S. aureus* [148–150]. A Swiss meat-processing plant reported the presence of *S. aureus* on 22.7% of the received chilled pork hind quarters from 18 European suppliers [151]. While investigating German pork processing units, Kastrup [139] determined a MRSA detection frequency of 6% on meat trimmings, 2% on processing equipment, and 5% on employees. Beneke et al. [152] obtained a similar detection rate in the processing area of a German abattoir. In an experimental setting, *S. aureus* at a contamination level of 5–7 log CFU/100 cm^{-2} was detectable on dry stainless steel for at least 96 h. In The Netherlands, de Jonge et al. [39] assessed the presence of MRSA in three meat-processing facilities and two institutional kitchens. MRSA was not isolated from any human nose or hand swabs, but 33% of the participants carried MSSA and only 14.3% of the meat samples were contaminated with MRSA. A Dutch study [138] which found that 46% of the retail meat samples, the majority of which came from a single retail shop and contained *S. aureus*, had a high degree of clonal relationship, indicating cross-transmission at some point during processing in the shop.

To pin point the exact source of contamination, it is necessary that the process of slaughtering be analyzed critically. Slaughter and meat processing involve several steps, any of which could introduce contamination with MRSA. Scalding, the first step in the slaughter process, is carried out at 60–62°C for 6–8 min in scalding tanks with rotating bars or through long scalding tanks [153] to loosen the hair from the carcass. An analysis of the effect of scalding on the quantity of coagulase-positive *S. aureus* (CPS) on pig carcasses in two Swiss abattoirs indicated variable data [154]. CPS, isolated from 96 to 100% of all carcasses, was reduced to 18 and 20% along the slaughter line after scalding from one abattoir, but in the second abattoir, it increased to 99% at the end of the line. Dehairing that follows scalding is another critical step that involves mechanical treatment of the carcass with rotating scrapers and rubber flails. This step has a potential to increase dissemination of porcine bacteria from mouth, nose, skin, and intestinal tract due to the accumulation of detritus in the machine. Singeing, which involves the exposure of the carcass for 10–15 s at 900°C, has been reported to decontaminate the surface of pig carcasses and lead to a 2.5- to 3-log reduction in total bacterial counts [153, 155].

Using a probabilistic model, Vossenkuhl et al. [156] found that a high MRSA prevalence at the beginning of the slaughter line was reduced to a low level at the end of the slaughter line (**Figure 1**). However, some investigators reported no effect of singeing on the microflora [157], whereas others have indicated that the reduction achieved by singeing is frequently reversed by polishing, that cleans intensively a skin surface [158–160]. Evisceration of the intestinal tract is another source of contamination with fecal bacteria on the surface of carcasses [154, 161]. To minimize the bacterial contamination, pig carcasses are usually chilled overnight using conventional single-stage chilling regimes, spray chilling, ice bank chilling in humid air at 2°C, and rapid or ultra-rapid chilling [162, 163]. Spescha et al. [154] reported a 77% decrease in the proportion of *S. aureus*-positive carcasses after chilling. Freeze chilling at temperatures of −10 to −25°C for 45–60 min, followed by chilling at 2°C for 23 h reduced *S. aureus* by 1 log CFU cm^{-2} on untrimmed carcasses [151]. It is clear from the published reports that handling of carcasses, proper maintenance of equipment, and personal hygiene play critical roles in the control and spread of *S. aureus* in the final end product.

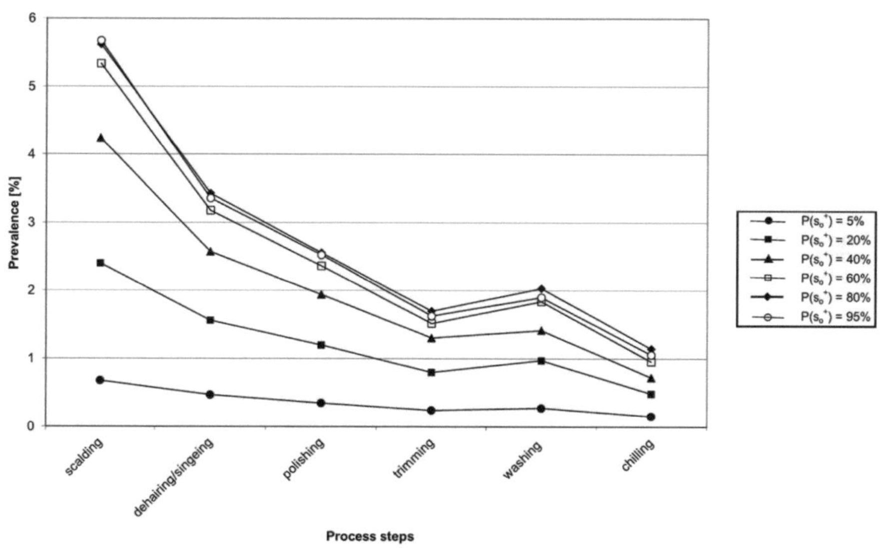

Figure 1. Change in MRSA prevalence along the slaughter line depending on the variation of the initial MRSA prevalence $P(s_0^+)$ [156]. Reproduced with permission of Elsevier.

4. Molecular epidemiology of *S. aureus* CC398

The emergence of LA-MRSA strains in humans [16, 17, 25, 56] and the presence of an identical MRSA CC398 in pigs, farm workers, veterinarians who attended to the same pig farms and their nonexposed family members [47, 48] suggests animal-to-human or human-to-animal transmission. In an interesting report from the Netherlands, it was shown that farm visitors were positive for CC398 MRSA directly after a farm visit but tested MRSA negative after

24 h [19, 41]. These and other studies indicate that CC398 appears to be frequently shared between animals and humans and is capable of causing infections in both species [25, 70, 164]. Transmission of MRSA between animals and humans is not new, but the MRSA isolates, in most cases, represent an initial human-to-animal transmission [24, 49, 165, 166].

Analysis of MRSA and MSSA from animals and humans spanning 19 countries and four continents indicated that the CC398 lineage originated in humans as an MSSA [167]. The whole-genome sequencing analysis by Price et al. [167] demonstrated that livestock-associated MRSA CC398 lost an immune-evasion cluster (IEC) as it evolved from its human-adapted MSSA. All of the HA-MSSA strains carry ØSa3 prophage in association with human innate immunomodulatory genes that play crucial roles in human niche adaptation (**Figure 2**) [167]. The prophage-associated virulence and adaptation genes are not necessary for nonhuman hosts, therefore, ØSa3 is mostly absent in livestock strains. After their introduction to livestock, MSSA CC398 acquired resistance to methicillin and tetracycline. Since tetracycline is heavily used in animal farming, the tetracycline resistance gene *tet*(M) is nearly universal among livestock-associated MRSA CC398 and MSSA isolates. The MSSA and MRSA CC398 isolates found in humans with direct livestock contact exhibited the same molecular patterns (i.e., ØSa3 prophage negative, *tet*(M)-positive) as the livestock-associated strains, indicating human re-adaptation [167], and were also reported in isolates epidemiologically associated with human-to-human transmission in multiple countries and continents [168]. During the host jump from humans to animals, MRSA CC398 strains also acquired resistance to copper and zinc because of their use in animal feed [130]. The vast majority of LA-MRSA CC398 strains carry SCC*mec* type Vc, which contains the *czrC* gene that confers resistance to copper and zinc.

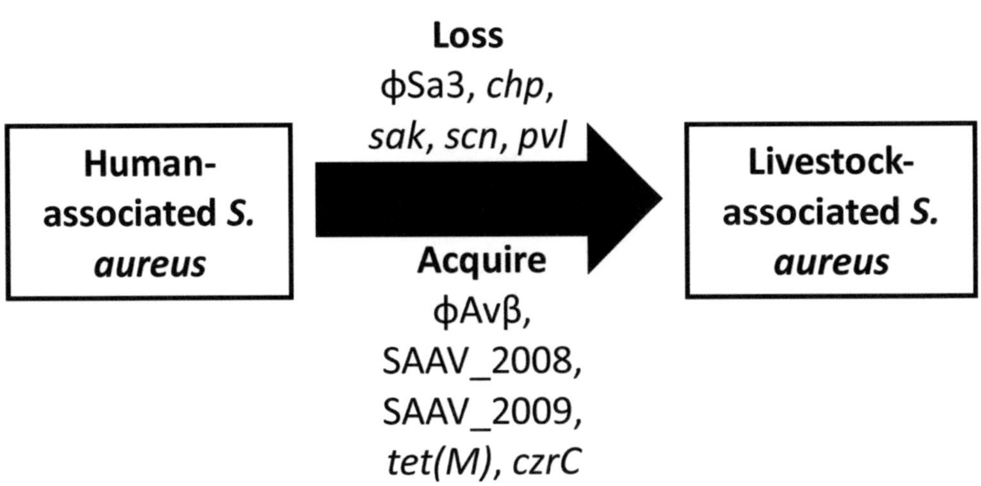

Figure 2. Gain or loss of genes as *S. aureus* jumps from human to livestock animals.

It is quite possible that LA-MRSA CC398 strains would eventually acquire certain genetic traits (additional antibiotic resistance and virulence factors) that would allow *S. aureus* to colonize both hosts and become a more formidable zoonotic agent [167]. Since ST398 strains are deficient in one or more restriction modification systems [169, 170], this adaptation process may have already occurred, as *pvl*-positive ST398 MRSA strains have been isolated from severe cases of community-acquired infections [20, 21]. These *pvl*-positive ST398 MRSA strains are also lysogenized with ϕSa3 phage coding for the human-specific virulence factors *sak, scn,* and *chp*. LA-MRSA *S. aureus* CC398 is not alone in the ability to adapt. A phylogenetic study of *S. aureus* CC97 strains that originated from livestock and caused human infections in Asia, Europe, and North and South America indicated that in the process of adapting to the human host, CC97 acquired either the SCC*mec* IV or V cassette and the ϕSa3 containing an immune evasion cluster [123]. A more recent study in the United States demonstrated that 22% of 30 veterinary students who were initially MRSA negative became positive after visiting MRSA-positive pig farms in Iowa but were negative again by 24 h after the visit [75]. The predominant *spa* type most commonly detected among the students was associated with ST5, suggesting a possible expansion of LA-MRSA to include ST5 [75].

Bayesian phylogenetic analysis of the poultry isolates with the CC5 clade indicated that it arose due to a single human-to-poultry host jump in or near Poland, where CC5 poultry strains acquired an MGE (mobile genetic element), presumably from other resident staphylococcal strains [36]. In addition to gene acquisition, loss of staphylococcal protein A (SpA) has occurred in avian isolates; it encodes virulence factors involved in human disease pathogenesis but is not needed for avian pathogenesis. The lack of protein A expression is a characteristic of the poultry biotype as defined by Devriese et al. [148]. In addition to the CC5 poultry clade, several other poultry isolates had identical or closely related STs to strains commonly associated with humans but had acquired MGE unique to avian strains, indicating that human-to-poultry host switches may be happening relatively frequently. A recent study demonstrated frequent contamination of poultry meat products with *S. aureus* ST5 and ST398 isolates, both of which have human origin [171]. Host adaptation of ST5 from humans to chickens was associated with a loss of genes contributing to human pathogenesis, and this was followed by the acquisition of avian-specific virulence determinants [36, 172]. Further, the presence of a cysteine protease encoded by a plasmid and widely distributed within avian strains suggests its potential role in avian-specific pathogenesis [173]. The adaptation from one host to another appears to be dependent upon the acquisition or loss of MGEs that code for key elements necessary for survival in new host [167, 174]. These cases have not only been associated with livestock [19] but have also occurred in cases with no known livestock contact [53, 54], suggesting a broader transmissibility capacity than originally thought. Further research is required to characterize the full scope of the genetic changes associated with the shift from humans to livestock and vice versa.

5. ST398 evolution and genetic diversity

In spite of similarities between LA- and HA-MRSA isolates, significant amounts of genetic diversity among *spa* and SCC*mec* cassette types have been documented in ST398 [6, 57, 175]. For instance, ST398 appears to have evolved by multiple acquisitions of the SCC*mec* elements, such as SCC*mec* types II, III, IV, IVa, and V [176]. In the Netherlands, two farms were found to have MRSA ST398 with identical *spa* types, but different SCC*mec* types, suggesting that divergent SCC*mec* elements were inserted into the clonal MSSA [57]. Similarly, MSSA ST398 (*spa* type t899), MRSA ST398–IVa (*spa* type t899), and MRSA ST398–V (*spa* type t108) were found in dust samples, nasal swabs, and a blood isolate from workers on the same pig farm [177], suggesting multiple acquisitions of SCC*mec* cassettes by MSSA precursors. Coagulase-negative staphylococci in the farming environment are suspected as sources of SCC*mec* [176], and the progeny of emerging MRSA strains are spreading locally rather than globally [178–180]. While SCC*mec* acquisition seems to be fairly common in MRSA ST398, the transfer of staphylococcal toxin genes, including the Panton-Valentine leukocidin gene (*pvl*) appears to be rarer [20, 43, 44, 57, 165, 181–184]. Only a handful of studies have found *pvl* positive ST398 [20, 165, 185–187]. Additionally, horizontal transfer of the protein A gene has been suggested, due to the finding of the *spa* type t899 in both ST398 strains and ST9 strains [29, 30, 80, 177].

6. MRSA in companion animals

6.1. Canine and feline MRSA

The first MRSA from pet animals was isolated from dogs in Nigeria in 1972 [188]. *S. intermedius* is the strain most isolated from dogs [189]. However, the predominant canine species among staphylococci was *S. sciuri* in Japan [190]. Various coagulase-positive and -negative staphylococci have been reported in pet animals [191]. Among coagulase-negative staphylococci, *S. felis* was dominant in Brazil [116], and a coagulase-positive *S. intermedius* was dominant in canine species in the UK [192, 193]. The prevalence of canine MRSA was 0.7% in Portugal, 2.3–9% in UK, and ≤20% in Canada [194–197]. Conversely, in cats, it was 1.48% in the UK, ≤4% in Portugal [195, 198] and 21.4% from wounds and skin lesions of cats in the USA [199]. *S. aureus* was found in 8% of dogs with inflammatory skin disease in the USA and one isolate was MRSA [200]. In addition, one MRSA was detected among the *S. aureus* strains isolated from 29% of 48 cats suffering with inflammatory skin disease and two MRSA were found among the *S. aureus* strains from 20% of 50 healthy cats [201]. Previous surgery, hospitalization, antimicrobial agent treatment, contact with humans possessing MRSA, and use of implant devices are regarded as risk factors for MRSA infection in companion animals [195, 202]. Rich and Robert [203] reported that MRSA were isolated from 1.4% of the postoperative and wound infection samples of pet animals in the UK. Lilenbaum et al. found 3% MRSA in Brazilian cats [193]. Southwest Pacific clone-associated community-acquired MRSA (USA1100) and methicillin-resistant *S. pseudintermedius* (MRSP)-associated with the European clone (ST71) were first reported in South America in cats [204].

MRSA isolates from Austria, Belgium, Germany, Ireland, and Portugal were resistant to ciprofloxacin and enrofloxacin, perhaps because of the fluoroquinolone approval for use in companion animals in Europe in the middle of 1990 (**Table 2**). MRSA ST398 that was identified in dogs and cats in France carried a chloramphenicol acetyltransferase gene, *cat*, that mediates resistance to nonfluorinated phenicols [117]. On the other hand, MRSA isolated in Portugal and Thailand from dogs and Germany from cats possessed a florfenicol-chloramphenicol exporter gene, *fexA*, that mediates resistance to both fluorinated and nonfluorinated phenicols [205–207]. MRSA ST398 isolated from an Austrian dog suffering from vaginitis harbored the *vatC* gene that inactivates streptogramin A [208]. An ABC transporter gene, *lsaE* (responsible for combined resistance to lincosamides, pleuromutilins, and streptogramin A), and a lincosamide nucleotidyltransferase *lunB* gene (confers resistance only to lincosamides), were present in the MRSA ST45 isolated from dogs in Thailand [205]. In France, an MRSA strain (*agr* III-t008–IV) isolated from a synovial fluid of dog was positive for *pvl* [117]. It carried *sek* and *seq* genes and was suspected to be a USA300 variant that was imported from the USA. *tsst1*-positive MRSA (*agr* II-t002-SCC*mec* I-truncated) was recovered from pet animals in France (**Table 3**). Leukotoxin genes, most prevalently *lukF* and/or *lukS* followed by *lukD* and/or *lukE*, are reported to be present in most of the MRSA isolated from companion animals, but those from dogs, cats and horses in Austria and the USA did not possess any leukotoxin genes. MRSA isolates CC5-t002–II, CC5-t062 and CC22-t032–IV, isolated in Portugal from a dog, a horse and human, respectively, harbored IEC genes suggesting adaptation to different hosts [206]. The presence of host-adaptive, virulence, and toxin genes makes the companion animal MRSA isolates prime candidates for zoonotic transfer. The first outbreak of human MRSA associated with cats was reported in a rehabilitation geriatric ward in the UK in 1988 [209] but others have since been reported in Australia, Canada, Germany, New Zealand, South America, and the Netherlands [193, 210–212]. Human-associated MRSA, such as EMRSA-15 and CMRSA-2, show a close relationship with pet-associated MRSA [211–214]. Since MRSA ST398 has been isolated from many farm animals, this MLST type found in dogs or cats may have originated from them or people who had contact with them [215]. MRSA SCC*mec* II strains, which are those most frequently affiliated with nosocomial human infections, have been isolated from cats [216].

6.2. Equine MRSA

Since the first report of MRSA from mares with metritis in Japan [217], many isolates have been found in Europe, North America, and Asia (**Table 2**) [142–144, 218–220]). Haenni et al. [221] identified four *mecC*-positive MRSA from horses in France exhibiting *spa* types, t208, t843, t6220, t11015 and ST types ST49, ST130, and ST1245. Among them, MRSA CC130 (ST130, ST1245) and CC49 (ST49) were documented in animals and humans, respectively. Multidrug-resistant ST8 MRSA was detected in Belgium, Germany, Switzerland, and the USA. A single locus variant of ST8 classified as ST254 MRSA was isolated from horses in Austria, Germany, Ireland, and the UK [183, 208, 222, 223]. Twelve different MLST types have been reported and most of the MRSA strains were grouped into the CC8 or CC398 classes [224]. The CC8-SCC*mec* IV genotype in horses was likely from a contaminated veterinary hospital and later spread to various clinics [219]. This genotype was reported in veterinary hospitals in Canada and the USA [118, 225]. It was first found in infected horses in Ireland and thereafter reported in the Netherlands, Austria, and Germany [226–228]. Many MRSA isolates from Canada showed ST8-SCC*mec* IV-t064 genotype [93]. They were designated as a Canadian epidemic MRSA-5 (CMRSA-5) and were very close to a human clone, USA500. Although the CC8-SCC*mec* IV genotype has been the most frequently found, CC398-SCC*mec* IV has recently become a major genotype. CC398-SCC*mec* IV was first found from infected horses in the Veterinary University of Vienna in Austria [222]. CC398 isolated from nasal samples of horses in the Netherlands and Belgium exhibited high prevalence rates, 9.3 and 10.9%, respectively [227, 229]. More than 25 *spa*-types have been reported, and three types, such as t011, t064 and t451, were the most

Country	Animal	ST	AMP	CEF	CHL	CIP	CLI	ENR	ERY	FLO	FUS	GEN	KAN	LIN	LIZ	QD	SPE	STR	TET	TIA	TOB	TRI	SXT	Genotype	References
Belgium	B†	398	*			+**			+			+	+	+				+	+		+	+			[87]
Belgium	B	398			+	+			+			+	+	+					+		+	+			[133]
Belgium	C†	239, 398		+	+		+		+		+	+	+			+		+	+	+		+			[34]
Belgium	C	398				+			+					+					+			+			[133]
Belgium	P†	ST9, 80, 239, 398			+	+	+		+		+	+	+		+	+		+	+	+		+	+	*aacA-aphD, aadD, aphA3-sat, blaI, blaR, blaZ, cfr, dfrS1, ermB, ermC, fexA, fosB, lnuA, tetK, tetM, vgaA*	[121]
Belgium	P	398				+			+			+	+	+					+		+	+			[256]
Belgium	P	398			+	+			+		+	+	+	+					+		+	+			[66]
Belgium	P	398			+	+			+			+	+	+		+			+		+	+			[133]
Brazil	B	398		+		+	+					+						+	+		+			*ant4, aac(6')-aph(2"), blaZ, mecA, tetK, tetM*	[257]
Central Europe	P	9			+		+		+										+				+		[258]

Country	Animal	ST	AMP	CEF	CHL	CIP	CLI	ENR	ERY	FLO	FUS	GEN	KAN	LIN	LIZ	QD	SPE	STR	TET	TIA	TOB	TRI	SXT	Genotype	References
China	P	6, 9, 63, 627	+	+	+		+		+	+					+		+		+	+				*aadE, erm33, fexA, lnuB, lsaE, mecA, spw, tetL, vgaA*	[259]
China	P	5, 9			+	+	+		+			+							+				+		[260]
China	P	NA		+	+	+	+		+			+							+				+	*acc(6')/ aph2", ant(4', 4"), aph(3')-III, ermC, mrsA, tetK, tetM*	[261]
China	P	398		+	+	+	+		+			+				+			+				+		[262]
Finland	P	1, 398		+		+	+		+										+					*aadD, blaZ/I/R, ermB, lnuB, mecA, tetK, tetK/M*	[120]
Germany	B	398	+		+		+		+			+	+	+		+	+		+	+		+	+	*aacA-aphD, aadD, aphA3, blaZ-I-R, dfrK, ermA, ermB, ermC, ermT, fex, spc, tetL, tetK, tetM, vgaA, vgaC*	[97]

Country	Animal	ST	AMP	CEF	CHL	CIP	CLI	ENR	ERY	FLO	FUS	GEN	KAN	LIN	LIZ	QD	SPE	STR	TET	TIA	TOB	TRI	SXT	Genotype	References
Germany	B	398	+					+	+										+						[102]
Germany	P	398	+	+	+		+	+	+	+		+	+				+		+	+			+	*aacA-aphD, aadD, blaZ, dfrG, dfrK, ermA, ermB, ermC, fexA, mecA, qacC, tetK, tetL, tetM, vga vgaA*	[44]
Hong Kong	C	9	+	+	+		+	+	+	+		+	+	+		+	+	+	+	+		+	+	*aacA-aphD, aadD, aadE, blaZ/I/R, dfrK, ermC, lnuB, lsaE, mecA, spw, tetL*	[134]
Hong Kong	P	9			+	+	+		+			+							+					*ermB, ermC*	[81]
Hong Kong	P	9	+	+			+	+				+	+				+	+	+	+		+		*aacA-aphD, aadD, aadE, blaZ/I/R, dfrK, ermC, lnuB, lsaE, mecA, tetL, spw*	[134]

http://dx.doi.org/10.5772/66645

Country	Animal	ST	AMP	CEF	CHL	CIP	CLI	ENR	ERY	FLO	FUS	GEN	KAN	LIN	LIZ	QD	SPE	STR	TET	TIA	TOB	TRI	SXT	Genotype	References
Hong Kong	P	9			+	+	+		+		+					+			+				+	*ermC, fexA, tetK, tetM*	[263]
Hungary	B	1	+						+										+						[92]
Ireland	P	CC398	+		+				+			+	+	+			+		+		+	+		*aacA-aphD, aadD, blaZ, dfrG, dfrK, ermA, ermB, fexA, spc, tetK, tetL, tetM*	[264]
Italy	B	CC97		+		+	+		+			+	+			+		+	+	+		+		*aacA-aphD, aadD, blaI, blaR, blaZ, cat, ermB, ermC, mecA, qacC, sdrM, tetK, tetM, vgaA*	[110]
Italy	B	1		+		+			+			+	+					+	+	+		+		*aphA3, blaZ, ermC, mecA, sat, tetK, sdrM*	[119]
Italy	P	1, 9, 97, 398, 1476			+	+	+		+			+	+					+	+			+			[62]

Country	Animal	ST	AMP	CEF	CHL	CIP	CLI	ENR	ERY	FLO	FUS	GEN	KAN	LIN	LIZ	QD	SPE	STR	TET	TIA	TOB	TRI	SXT	Genotype	References
Italy	P	CC97		+		+	+		+			+	+			+		+	+	+		+		*aacA-aphD, aadD, blaI, blaR, blaZ, cat, ermB, ermC, mecA, qacC, sdrM, tetK, tetM, vgaA*	[110]
Italy	P	1		+		+	+		+			+	+					+	+	+			+	*aacA-aphD, aadD, blaI, blaR, blaZ, cfr, dfrS1, ermA, ermC, ermR, fexA, mecA, qacC, sdrM, tetK, tetM, vgaA*	[119]
Italy & Spain	P	1	+			+	+		+			+	+						+	+		+		*aacA-aphD, aadD, blaZ, cat, dfrA, ermA, ermC, lnuA, mecA, tetK, tetL, tetM, vgaA, vgbA*	[265]

Country	Animal	ST	AMP	CEF	CHL	CIP	CLI	ENR	ERY	FLO	FUS	GEN	KAN	LIN	LIZ	QD	SPE	STR	TET	TIA	TOB	TRI	SXT	Genotype	References
Korea	B	NA	+	+		+	+		+			+	+						+						[89]
Korea	B	1, 72							+			+							+						[108]
Korea	B	NA	+				+		+			+	+						+						[90]
Korea	C	692				+	+		+										+					*ermB, ermC, tetL, tetS*	[91]
Korea	C	NA	+	+			+		+																[89]
Korea	P	398				+	+		+										+						[85]
Korea	P	5, 72, 398, 541			+	+	+		+		+	+	+		+	+			+	+				*aac(6')-le-aph(2"), ant(3')-IIIa, anat(4')-la, cfr, ermA, ermC, ermT, fexA, lnuB, tetK, tetL, tetM*	[91]
Netherland	B	398	+									+							+		+		+		[266]
Netherlands	P	398					+		+			+	+						+		+				[22]
Netherlands	P	398							+			+	+	+					+						[70]
Portugal	B	398			+	+	+	+	+	+			+						+					*aphA3, ermC, fexA, tetK, tetM*	[206]

Country	Animal	ST	AMP	CEF	CHL	CIP	CLI	ENR	ERY	FLO	FUS	GEN	KAN	LIN	LIZ	QD	SPE	STR	TET	TIA	TOB	TRI	SXT	Genotype	References
Portugal	P	NA				+	+		+			+	+						+	+		+	+	*aacA-aphD, aadD, apmA, dfrK, ermC, fexA, qacG, qacJ, tetK, tetM, vgaA*	[206]
Spain	P	398, 1379					+		+			+	+					+	+		+		+	*aacA-aphD, aadA, aadD, aphA3, dfrA, dfrG, dfrK, ermC, ermT, msrA, str, tetK, tetM, tetL*	[255]
Spain	P	398					+		+			+									+		+		[234]
Spain	P	1, 398, 1965, 1966, 1967, 1968, 1969		+	+	+			+		+	+		+		+		+	+			+			[267]
Switzerland	B	1, 398	+	+			+		+										+					*aphA,blaI, blaR, blaZ, ermC, mecA, tetM*	[42]

http://dx.doi.org/10.5772/66645

Country	Animal	ST	AMP	CEF	CHL	CIP	CLI	ENR	ERY	FLO	FUS	GEN	KAN	LIN	LIZ	QD	SPE	STR	TET	TIA	TOB	TRI	SXT	Genotype	References
Switzerland	P	398	+	+			+		+			+							+				+	*blaI, blaR, blaZ, ermA, mecA, tetK, tetM*	[42]
Switzerland	P	398		+			+		+					+				+	+						[63]
Taiwan	P	9			+	+	+		+			+							+				+		[268]
Thailand	P	9		+	+		+					+							+				+		[31]
Thailand	P	9, 2136		+		+						+					+		+	+		+	+	*blaZ, aac-aphD, tetM, vgaA*	[269]
USA	P	398					+		+							+			+						[51]
USA	P	5, 72, 398, 1340	+				+		+			+				+			+				+		[38]

B†: cattle, C†: chicken, P†: pigs, *: not tested or sensitive, **: resistant, AMP: ampicillin, CEF: cefoxitin, CHL: chloramphenicol, CIP: ciprofloxacin, CLI: clindamycin, ENR: enrofloxacin, ERY: erythromycin, FLO: florfenicol, FUS: fusidate, GEN: gentamicin, KAN: kanamycin, LIN: lincomycin, LIZ: linezolid, QD: quinupristin/dalfopristin, SPE: spectinomycin, STR: streptomycin, TET: tetracycline, TIA: tiamulin, TOB: tobramycin, TRI: trimethoprim, SXT: trimethoprim-sulfamethoxazole.

Table 1. Antimicrobial resistance pattern of bovine, poultry, and porcine MRSA isolates.

Country	Animal	ST	AMP	CEF	CHL	CIP	CLI	ENR	ERY	FLO	FUS	GEN	KAN	LIN	LIZ	QD	SPE	STR	TET	TIA	TOB	TRI	SXT	Genotype	References
Austria	D†	22, 254, 398	*			+**			+			+							+				+	*aacA-aphD*, *dfrA*, *ermC*, *tetK*, *tetM*, *vatC*	[208]
Austria	C†	1, 5, 398				+	+		+			+							+					*aacA-aphD*, *ermC*, *tetK*, *tetM*	[208]
Austria	H†	1, 254, 398				+	+		+			+							+				+	*aacA-aphD*, *dfrA*, *ermC*, *tetK*, *tetM*, *vatC*	[208]
Austria	H	1, 254, 398					+		+			+							+					*aph2″-aac6′*, *ermC*, *mecA*, *tetM*	[222]
Belgium	D	398			+	+			+			+	+	+					+		+	+			[133]
Belgium	C	398			+	+			+					+		+			+						[133]
Belgium	H	8, 398, 2197	+	+			+	+	+			+		+					+			+			[270]
Belgium	H	398					+		+			+		+					+			+			[229]
Brazil	C	30		+																					[204]
Brazil	C	NA	+																						[193]
Canada	D	NA	+			+	+	+	+		+	+							+				+		[271]
Canada & USA	H	NA			+			+	+			+							+			+			[272]
China	C, D	59, 398					+					+							+					*aacA-aphD*, *ermB*, *mecA*, *linA*, *tetK*	[273]

Country	Animal	ST	AMP	CEF	CHL	CIP	CLI	ENR	ERY	FLO	FUS	GEN	KAN	LIN	LIZ	QD	SPE	STR	TET	TIA	TOB	TRI	SXT	Genotype	References
France	C, D	CC398		+	+			+	+		+		+	+					+		+			*aadD, blaZ, cat, dfrA, erm(A), fosB, mecA, qacA, tetK, tetM, vgaA*	[117]
Germany	C, D, H	CC5, 8, 9, 22, 398, 599			+		+	+	+			+	+						+				+		[274]
Germany	C	398	+	+	+		+		+					+		+	+		+				+	*blaZ, dfrK, ermC, fexA, mecA, tetK, tetM*	[207]
Germany	C, D	22			+	+	+		+															*ermC, mecA*	[210]
Germany	H	8, 22, 254, 1117																						*aacA-aphD, aadD, aphA3, blaI/R, blaZ, cat, dfrA, ermA, ermC, fosB, mecA, qacA, sat, tetK, tetM*	[183]
Hong Kong	D	NA			+	+	+		+		+	+							+				+		[275]
Ireland	D	NA	+			+			+					+					+						[276]
Ireland	D	NA	+			+			+					+					+						[226]
Ireland	H	NA	+			+			+			+	+					+	+		+	+			[226]
Israel	H	5				+	+		+			+													[277]
Israel	H	5				+	+		+			+													[278]

Country	Animal	ST	AMP	CEF	CHL	CIP	CLI	ENR	ERY	FLO	FUS	GEN	KAN	LIN	LIZ	QD	SPE	STR	TET	TIA	TOB	TRI	SXT	Genotype	References
Japan	D	5, 30				+			+			+	+					+	+						[279]
Japan	D	NA	+						+			+							+						[280]
Korea	D	72		+																					[281]
Korea	D	NA	+		+	+		+				+									+		+		[282]
Malaysia	D	59		+					+									+	+						[283]
Malaysia	C	55		+					+			+						+	+						[283]
Netherlands	D	NA	+					+	+			+		+											[284]
Nigeria	D	NA		+	+	+	+		+			+							+				+		[285]
Portugal	D	22, 105, 398			+	+	+		+	+			+						+					*fexA*, *ermA*, *ermC*, *tetM*	[206]
Portugal	C	5, 22				+	+		+		+													*ermC*, *fusC*, *mphC*, *msrA*,	[206]
Portugal	D	22				+	+		+		+	+	+			+			+		+			*aac(6')-aph(2")*, *ant(4')-Ia*, *aph(3')-III*, *ermB*, *ermC*, *msrA*, *tetM*	[286]
Portugal	H	5, 398					+		+		+	+	+						+			+		*aacA-aphD*, *blaZ*, *dfrK*, *ermC*, *fusC*, *tetM*	[287]

Country	Animal	ST	AMP	CEF	CHL	CIP	CLI	ENR	ERY	FLO	FUS	GEN	KAN	LIN	LIZ	QD	SPE	STR	TET	TIA	TOB	TRI	SXT	Genotype	References
Portugal	H	5					+		+		+	+	+						+			+	+	*aacA-aphD, dfrK, ermC, fusC, tetM*	[206]
Switzerland	H	8, 398				+	+		+			+	+					+	+			+		*aac(6')-le-aph(2')-la, ant(6)-Ia, aph(3')-III, blaZ, dfrG, dfrK, ermC, mecA, mphC, msrA, str, tetK, tetM*	[288]
Thailand	D	45			+	+	+					+	+			+		+	+	+		+		*aac(6')-Ie-aph(2')-Ia, ant(4')-Ia, ant(6')-Ia, blaZ, dfrA, fexA, lnuB, lsaE, mecA, mupR, tetL, tetM*	[205]
USA	D	5	+			+	+		+															*blaZ, ermA, mecA*	[289]
USA	D	NA	+			+	+	+	+			+							+				+		[290]
USA	D	5, 8, 105, 986	+					+	+			+							+						[118]
USA	C	5, 8	+					+	+										+						[118]
USA	C	NA	+	+			+	+	+			+													[291]
USA	D	NA	+																						[291]

Country	Animal	ST	AMP	CEF	CHL	CIP	CLI	ENR	ERY	FLO	FUS	GEN	KAN	LIN	LIZ	QD	SPE	STR	TET	TIA	TOB	TRI	SXT	Genotype	References
USA	C, D	NA			+		+	+	+			+							+				+		[292]
USA	D	NA				+	+		+										+						[293]
USA	C	NA				+	+		+										+						[293]
USA	C	NA			+		+	+	+			+							+				+		[216]
USA & UK	D	NA				+	+	+	+			+		+									+		[294]
USA	H	8, 830	+						+			+							+				+		[118]

$C^{\dagger}$: cats, $D^{\dagger}$: dogs, $H^{\dagger}$: horses, *: not tested or sensitive, **: resistant, AMP: ampicillin, CEF: cefoxitin, CHL: chloramphenicol, CIP: ciprofloxacin, CLI: clindamycin, ENR: enrofloxacin, ERY: erythromycin, FLO: florfenicol, FUS: fusidate, GEN: gentamicin, KAN: kanamycin, LIN: lincomycin, LIZ: linezolid, QD: quinupristin/dalfopristin, SPE: spectinomycin, STR: streptomycin, TET: tetracycline, TIA: tiamulin, TOB: tobramycin, TRI: trimethoprim, SXT: trimethoprim-sulfamethoxazole.

Table 2. Antimicrobial resistance pattern of feline, canine and equine MRSA isolates.

http://dx.doi.org/10.5772/66645

Country	Animals	Leukotoxins	Hemolysins	Enterotoxins	Immune-invasion factors	Protease	Superantigen-like proteins	Biofilm-associated	Capsule	Miscellaneous genes	References
Austria	Horses			*seb*							[208]
	Dogs			*sei*							
	Cats			*sei*							
Belgium	Pigs	*lukD/E, lukF, lukPV, lukS, lukX, lukY*	*hla, hlb, hld, hlIII, hlgA*	*seg, sei, sem, sen, seo, seu*	*hysA1, hysA2, isaB, isdA*	*splA, splB*	*ssl1, ssl2, ssl3, ssl4, ssl5, ssl6, ssl7, ssl8, ssl9, ssl10, ssl11*	*bbp, cflA, cflB, cna, ebh, ebpS, eno, fib, fnbA, fnbB, icaACD, map, mprF, sasG, sdrC, sdrD, vwb*	*capH5, capH8, capI8, capJ5, capJ8, capK5, capK8*	*edinB, etd, setC*	[121]
China	Pigs			*sec, seg, sei, sem, sen, seo, seh*							[260]
Finland	Pigs			*seh*				*fnbB*	*cap5, cap8*		[120]
France	Cats and dogs	*lukF-PV, lukS-PV*		*sea, sec, sed, seg, sei, sej, sek, sel, sem, sen, seo, seq, ser, seu*						*tst*	[117]
Germany	Pigs			*seb, sek, seq*							[44]
	Horses	*lukD/E, lukF, lukS*	*hl, hla, hlb, hld, hlIII-aII, hlgA*	*sea, seb/k/g, sec/l, sed/j/r, seg/i/m/n/o/u, sep*	*chp, sak, scn*	*splAB/E, splE, sspA/B/P*					[183]

Country	Animals	Leukotoxins	Hemolysins	Enterotoxins	Immune-invasion factors	Protease	Superantigen-like proteins	Biofilm-associated	Capsule	Miscellaneous genes	References
Italy	Cattle	*lukD, lukE, lukF, lukS, lukX, lukY*	*hl, hlIII, hlb, hlgA*	*seh*	*hysA1, hysA2, isaB, isdA, sak, scn*	*aur, splA, splB, sspA, sspB, sspP*	*ssl1, ssl2, ssl3, ssl4, ssl5, ssl6, ssl7, ssl8, ssl9, ssl10, ssl11*	*bbp, clfA, clfB, cna, ebh, ebpS, icaA, icaC, icaD, eno, fib, fnbA, fnbB, sasG, sdrC, sdrD, vwb*	*cap8, capH8, capI8, capJ8, capK8*	*lmrP, mprF, hsdS2, hsdS3, hsdSx*	[119]
	Pigs	*lukD, lukE, lukF, lukS, lukX, lukY*	*hl, hla, hlIII, hlb, hlgA*	*seh*	*isaB, isdA, hysA1, hysA2*	*aur, splA, splB, splE, sspA, sspB, sspP*	*ssl1, ssl2, ssl3, ssl4, ssl5, ssl6, ssl7, ssl8, ssl9, ssl10, ssl11*	*bbp, clfA, clfB, cna, ebh, ebpS, icaA, icaC, icaD, eno, fib, fnbA, fnbB, sasG, sdrC, sdrD, vwb*	*cap8, capH8, capI8, capJ8, capK8*	*lmrP, mprF, hsdS2, hsdS3, hsdSx*	
Italy & Spain	Pigs	*lukX*	*hlb*	*entH, entU, entX, entY*							[265]
Korea	Pigs	*lukE/D*		*sed, seg, sei, sej, sem, sen, seo, sep, ser*	*chp, sak, scn*					*tsst1*	[91]
	Chicken	*lukE/D*		–	–						
	Cattle	*lukE/D, pvl*		*seb, sed, seg, sei, sej, sek, sem, sen, seo, seq, ser*	*chp, sak, scn*					*tsst1*	

http://dx.doi.org/10.5772/66645

Country	Animals	Leukotoxins	Hemolysins	Enterotoxins	Immune-invasion factors	Protease	Superantigen-like proteins	Biofilm-associated	Capsule	Miscellaneous genes	References
Portugal	Pigs	*lukF, lukS*	*hla, hld*	–	–						[206]
	Dog	*lukD, lukE, lukF, lukS,*	*hla, hlb, hld*	*sec, sed, seg, sei, sej, sel, sem, sen, seo, seu, ser*	*chp, sak, scn*						
	Horse	*lukF, lukS, lukD, lukE*	*hla, hlb, hld*	*seg, sei, sem, sen, seo, seu*	*chp, sak, scn*						
	Calf	*lukF, lukS*	*hla, hld*	–	–						
Spain	Pigs	*lukD, lukE*	*hla, hlb, hld, hlg, hlg-v*							*eta*	[255]
Switzerland	Pigs	*lukF, lukS, lukY*	*hla, hld, hlgA*	*entX, entY*							[42]
	Calf	*lukF, lukS, lukY*	*hla, hld, hlgA*	*entX*							
	Cattle	*lukD, lukE, lukF, lukY*	*hla, hld, hlgA*	*entH, entX, entY*							
Thailand	Pigs			*entG, entI, entM, entN, entO*							[269]
USA	Dogs	*lukSF-PV*		*sea, seb, sec, sed, seg, sei, sej, sem, sen*				*clfA, clfB, fnbA, fnbB*			[118]
	Cats	*lukSF-PV*		*sec, sed, seg, sei, sej, sek, sem, sen*				*clfA, clfB, fnbA, fnbB*			
	Horses			*sea, seb, seg, sek, sei, sej, sek*				*clfA, clfB, fnbA, fnbB*			

Table 3. Virulence profiles from food-producing and pet animals.

widespread [230]. In addition, only three SCC*mec* types (IV, V, VI) have been discovered from horses [218]. Interestingly, no MRSA was isolated from 300 horses on 14 farms in Slovenia, 497 horses on 50 farms in Canada, 87 horses in Austria, and 200 horses in the Netherlands [231–233]. ST22 and ST1117 isolates from horses in Germany had IEC genes, such as *chp, sak,* and *scn* [183]. On the other hand, ST8, ST254, and ST398 strains did not carry those genes.

6.3. Antibiotic resistance and enterotoxin genes in LA-MRSA CC398

An analysis of MRSA and MSSA from animals and humans spanning 19 countries and four continents indicated that the CC398 lineage originated in humans as MSSA [167]. After its transmission to livestock, CC398 became resistant to tetracycline, probably because of the heavy tetracycline use in pig production [22]. However, many tetracycline-resistant MRSA strains are found in horses despite the fact that tetracycline is either not used much [229] or sparingly used [93, 229]. Among bovine MRSA isolates tested, most of them were resistant to β-lactam antibiotics [34] as well as tetracycline, erythromycin and gentamicin. CC398 has also been reported to be highly resistant to several other antibiotics, such as ciprofloxacin, tobramycin, clindamycin, and trimethoprim-sulfamethoxazole [234]. Antimicrobial resistance patterns of MRSA and MSSA isolates in Hong Kong were very similar [81]. The only MRSA CC398 isolate that has exhibited resistance against daptomycin and intermediate susceptibility to vancomycin has been described in a case-report from an Italian hospital [235]. Two other isolates, one from a ventilator-associated pneumonia of a farmer and one additional porcine isolate, were also described as resistant to linezolid and possessed the *cfr* gene which is located on transferable plasmids [235]. An outbreak with HA-MRSA ST125 containing *cfr*, reported from a Madrid hospital in 2010 [236], was followed by reports on the emergence of nosocomial coagulase negative staphylococci containing *cfr* [40]. This gene was previously found in staphylococci from animals in Europe [237, 238] but has also been reported in humans in Colombia and the USA [239, 240]. Distribution of the *cfr* gene in MRSA isolates is alarming because, besides linezolid, it confers resistance to oxazolidinones, phenicols, lincosamides, streptogramin A, and pleuromutilins, including the topical antibiotic retapamulin that is used for treatment of human skin infections [239, 241, 242].

About 50% of LA-MRSA CC398 isolates, besides being resistant to antimicrobial agents, also exhibit resistance to copper and zinc mediated by the *czrC* gene [130, 243]. The use of zinc as feed additives may have favored spread of the *czrC* gene in LA-MRSA [130]. So far, this gene has only been found in LA-MRSA [130, 243] but an extended use of copper coating of biomaterials in orthopedic surgery and traumatology, as well as copper surfaces in hospitals, might select for *czrC*-positive HA-MRSA as well. The trimethoprim-resistance gene *dfrK*, located close to *tetL* in LA-MRSA ST398, has also been found on a plasmid [244]. The *tetL* gene was identified in MRSA isolated from diverse livestock animals and meats from different regions of the world [97, 245, 246]. Kadlec and Schwarz demonstrated the presence of plasmid pKKS25-associated resistance genes, *ermT, dfrK,* and *tetL*, in MRSA obtained from a nasal swab of a young sow in Germany [247]. A porcine MRSA ST398 was shown to contain a transposon Tn*6133* carrying *ant(9)-Ia* and *ermA* genes and a plasmid pKKS825 [248, 249] harboring the resistance genes *vgaC* and *vgaE, aadD, tetL,* and *dfrK* [248–250]. In Germany, the *vgaE* gene was detected in MRSA ST398 isolated from cattle, turkeys, and chicken and turkey meats

[251]. An apramycin resistance gene, *apmA*, was discovered in a bovine MRSA ST398 [252] and in porcine MRSA ST398 [206, 253]. In Denmark, the quaternary ammonium compound-resistant genes, *qacC* and *qacG* were detected in MRSA CC30 isolates [254]. Wendlandt et al. identified a plasmid pV7037-associated multidrug resistance gene cluster, including the novel resistance genes *lsaE* and *spw*, and other resistance genes, such as *mecA*, *blaZ/I/R*, *tetL*, *dfrK*, *ermC*, and *aadD*, from frozen or chilled chicken carcasses in Hong Kong. The antimicrobial resistance patterns and associated genes from bovine, porcine, and poultry isolates are summarized in **Table 1** and those from companion animals are shown in **Table 2**.

Most of the animal isolates are negative for the *pvl* gene, but Belgian MRSA ST80–IV isolates from healthy pigs were positive for the *pvl* gene and corresponded to the community-acquired CA-MRSA ST80–IV European clone [121]. One Belgian pig CC80 strain contained an exfoliatin (*etd*), an epidermal cell differentiation inhibitor (*edinB*), and staphylococcal exotoxin-like protein (*setC*) and IEC genes (*isaB, isdA, hysA1, hysA2*) [121]. A comparison of the *pvl*-positive MRSA isolates (ST8-t008–IVa) from American pig and pet animals indicated the presence of common virulence profiles (*lukSF-PV, clfA, clfB, fnbA*, and *sek*) except for the *fnbB* gene of a canine isolate [118]. Spanish ST1379/CC97 porcine isolates carried an exfoliatin (*eta*), a leukotoxin (*lukE/D*), and a gamma-hemolysin (*hlg-2*) but were negative for *etb, etc, tst, pvl*, and enterotoxins [255]. The occurrence and prevalence of enterotoxin genes in MRSA isolates from food-producing or companion animals is summarized in **Table 3**. The presence of multidrug-resistant MRSA in companion and food animals, combined with the enterotoxin genes, will require constant monitoring and evaluation of mitigation strategies.

7. Conclusions

MRSA contamination in food-producing and companion animals poses a serious threat to public health. Incidences of identical LA-MRSA strains in pig farms and persons in close contact with food producing and companion animals suggest a clear link for transmission of these strains between humans and animals. While MRSA isolates from companion and food-producing animals are known to infect humans, the reverse is also true. Studies reviewed in this report indicate an initial transfer of MSSA from humans to animals by deletion of immunomodulatory genes and prophage ØSa3, necessary for human infection but not required for infection in animals, and acquisition of tetracycline and methicillin resistance genes (**Figure 2**). The MRSA that evolved in animals started showing up in humans that were in close contact with them and exhibited traits specifically found in animal isolates, indicating a reverse transmission from animals to humans. Initial reports of MRSA in animals did not indicate the presence of host adaptation, enterotoxin, virulence, and antimicrobial resistance genes in them but they are becoming more prevalent and it is feared that these animals could serve as a reservoir for such strains and play an important role in zoonotic transfers.

Documentation of MRSA isolates (ST59-t437–V) from cattle containing the *pvl* gene and other virulence factors, such as *chp, scn, seb, sek*, and *seq*, toxic shock syndrome toxin 1 (*tsst1* or *tst*) gene and hemolysin, protease, superantigen-like protein, capsule, and biofilm-associated genes make them powerful pathogens that could cause a medical nightmare.

A livestock-associated CC398 lineage MRSA is well known to transfer from animals to humans and other MRSA isolates of different clonal complex groups are also known to be associated with zoonotic transfers. A human pandemic community-associated CC97 lineage MRSA harboring the antimicrobial resistance genes *mecA* and *mecC* has been shown to have originated from animals.

A comprehensive study of the emergence, dissemination, prevention and control of MRSA colonization is required to mitigate the risks to both animal and human health. Rapid advancement of whole genome sequencing technology has the great power of discriminating closely associated MRSA isolates from different sources and could be used for source tracking and differentiating between animal and human origin isolates. In addition, it can be applied to monitor the emergence and dissemination of MRSA isolated from various environments and determine the characteristics of virulence factors and evolution of multi-antimicrobial resistance.

Disclaimer

The views expressed herein do not necessarily reflect those of the US Food and Drug Administration or the US Department of Health and Human Services.

Author details

Jungwhan Chon, Kidon Sung* and Saeed Khan

*Address all correspondence to: kidon.sung@fda.hhs.gov

Division of Microbiology, National Center for Toxicological Research, US Food and Drug Administration, Jefferson, AR, USA

References

[1] Hennekinne JA, De Buyser ML, Dragacci S. *Staphylococcus aureus* and its food poisoning toxins: characterization and outbreak investigation. FEMS Microbiol Rev. 2012;36(4):815–36.

[2] Weese JS. Methicillin-resistant *Staphylococcus aureus* in animals. ILAR J. 2010;51(3):233–44.

[3] Stryjewski ME, Corey GR. Methicillin-resistant *Staphylococcus aureus*: an evolving pathogen. Clin Infect Dis. 2014;58 Suppl 1:S10–9.

[4] Waness A. Revisiting methicillin-resistant *Staphylococcus aureus* infections. J Glob Infect Dis. 2010;2(1):49–56.

[5] Taylor AR. Methicillin-resistant *Staphylococcus aureus* infections. Prim Care. 2013;40(3):637–54.

[6] Pan A, Battisti A, Zoncada A, Bernieri F, Boldini M, Franco A, et al. Community-acquired methicillin-resistant *Staphylococcus aureus* ST398 infection, Italy. Emerg Infect Dis. 2009;15(5):845–7.

[7] McNamee PT, Smyth JA. Bacterial chondronecrosis with osteomyelitis ('femoral head necrosis') of broiler chickens: a review. Avian Pathol. 2000;29(5):477–95.

[8] Eriksen KR. ["Celbenin"-resistant staphylococci]. Ugeskr Laeger. 1961;123:384–6.

[9] Hiramatsu K, Katayama Y, Matsuo M, Sasaki T, Morimoto Y, Sekiguchi A, et al. Multi-drug-resistant *Staphylococcus aureus* and future chemotherapy. J Infect Chemother. 2014;20(10):593–601.

[10] Weber JT. Community-associated methicillin-resistant *Staphylococcus aureus*. Clin Infect Dis. 2005;41 Suppl 4:S269–72.

[11] Vanderhaeghen W, Hermans K, Haesebrouck F, Butaye P. Methicillin-resistant *Staphylococcus aureus* (MRSA) in food production animals. Epidemiol Infect. 2010;138(5):606–25.

[12] Pantosti A. Methicillin-resistant *Staphylococcus aureus* associated with animals and its relevance to human health. Front Microbiol. 2012;3:127.

[13] Bal AM, Coombs GW, Holden MT, Lindsay JA, Nimmo GR, Tattevin P, et al. Genomic insights into the emergence and spread of international clones of healthcare-, community- and livestock-associated meticillin-resistant *Staphylococcus aureus*: blurring of the traditional definitions. J Glob Antimicrob Resist. 2016;6:95–101.

[14] Fitzgerald JR. Livestock-associated *Staphylococcus aureus*: origin, evolution and public health threat. Trends Microbiol. 2012;20(4):192–8.

[15] Fluit AC. Livestock-associated *Staphylococcus aureus*. Clin Microbiol Infect. 2012;18(8):735–44.

[16] Wendlandt S, Kadlec K, Fessler AT, Monecke S, Ehricht R, van de Giessen AW, et al. Resistance phenotypes and genotypes of methicillin-resistant *Staphylococcus aureus* isolates from broiler chickens at slaughter and abattoir workers. J Antimicrob Chemother. 2013;68(11):2458–63.

[17] Armand-Lefevre L, Ruimy R, Andremont A. Clonal comparison of *Staphylococcus aureus* isolates from healthy pig farmers, human controls, and pigs. Emerg Infect Dis. 2005;11(5):711–4.

[18] Nemati M, Hermans K, Lipinska U, Denis O, Deplano A, Struelens M, et al. Antimicrobial resistance of old and recent *Staphylococcus aureus* isolates from poultry: first detection of livestock-associated methicillin-resistant strain ST398. Antimicrob Agents Chemother. 2008;52(10):3817–9.

[19] Graveland H, Wagenaar JA, Heesterbeek H, Mevius D, van Duijkeren E, Heederik D. Methicillin resistant *Staphylococcus aureus* ST398 in veal calf farming: human MRSA carriage related with animal antimicrobial usage and farm hygiene. PLoS One. 2010;5(6):e10990.

[20] Stegger M, Lindsay JA, Sorum M, Gould KA, Skov R. Genetic diversity in CC398 methicillin-resistant *Staphylococcus aureus* isolates of different geographical origin. Clin Microbiol Infect. 2010;16(7):1017–9.

[21] Sung JM, Lloyd DH, Lindsay JA. *Staphylococcus aureus* host specificity: comparative genomics of human versus animal isolates by multi-strain microarray. Microbiology. 2008;154(Pt 7):1949–59.

[22] de Neeling AJ, van den Broek MJ, Spalburg EC, van Santen-Verheuvel MG, Dam-Deisz WD, Boshuizen HC, et al. High prevalence of methicillin resistant *Staphylococcus aureus* in pigs. Vet Microbiol. 2007;122(3–4):366–72.

[23] van Loo I, Huijsdens X, Tiemersma E, de Neeling A, van de Sande-Bruinsma N, Beaujean D, et al. Emergence of methicillin-resistant *Staphylococcus aureus* of animal origin in humans. Emerg Infect Dis. 2007;13(12):1834–9.

[24] Duquette RA, Nuttall TJ. Methicillin-resistant *Staphylococcus aureus* in dogs and cats: an emerging problem? J Small Anim Pract. 2004;45(12):591–7.

[25] Smith TC, Pearson N. The emergence of *Staphylococcus aureus* ST398. Vector Borne Zoonotic Dis. 2011;11(4):327–39.

[26] Johnson AP. Methicillin-resistant *Staphylococcus aureus*: the European landscape. J Antimicrob Chemother. 2011;66 Suppl 4:iv43–iv8.

[27] Kock R, Mellmann A, Schaumburg F, Friedrich AW, Kipp F, Becker K. The epidemiology of methicillin-resistant *Staphylococcus aureus* (MRSA) in Germany. Dtsch Arztebl Int. 2011;108(45):761–7.

[28] Larsen J, Imanishi M, Hinjoy S, Tharavichitkul P, Duangsong K, Davis MF, et al. Methicillin-resistant *Staphylococcus aureus* ST9 in pigs in Thailand. PLoS One. 2012;7(2):e31245.

[29] Neela V, Mohd Zafrul A, Mariana NS, van Belkum A, Liew YK, Rad EG. Prevalence of ST9 methicillin-resistant *Staphylococcus aureus* among pigs and pig handlers in Malaysia. J Clin Microbiol. 2009;47(12):4138–40.

[30] Wagenaar JA, Yue H, Pritchard J, Broekhuizen-Stins M, Huijsdens X, Mevius DJ, et al. Unexpected sequence types in livestock associated methicillin-resistant *Staphylococcus aureus* (MRSA): MRSA ST9 and a single locus variant of ST9 in pig farming in China. Vet Microbiol. 2009;139(3–4):405–9.

[31] Patchanee P, Tadee P, Arjkumpa O, Love D, Chanachai K, Alter T, et al. Occurrence and characterization of livestock-associated methicillin-resistant *Staphylococcus aureus* in pig industries of northern Thailand. J Vet Sci. 2014;15(4):529–36.

[32] Fang HW, Chiang PH, Huang YC. Livestock-associated methicillin-resistant *Staphylococcus aureus* ST9 in pigs and related personnel in Taiwan. PLoS One. 2014;9(2):e88826.

[33] Argudin MA, Cariou N, Salandre O, Le Guennec J, Nemeghaire S, Butaye P. Genotyping and antimicrobial resistance of *Staphylococcus aureus* isolates from diseased turkeys. Avian Pathol. 2013;42(6):572–80.

[34] Nemeghaire S, Roelandt S, Argudin MA, Haesebrouck F, Butaye P. Characterization of methicillin-resistant *Staphylococcus aureus* from healthy carrier chickens. Avian Pathol. 2013;42(4):342–6.

[35] Buyukcangaz E, Velasco V, Sherwood JS, Stepan RM, Koslofsky RJ, Logue CM. Molecular typing of *Staphylococcus aureus* and methicillin-resistant *S. aureus* (MRSA) isolated from animals and retail meat in North Dakota, United States. Foodborne Pathog Dis. 2013;10(7):608–17.

[36] Lowder BV, Guinane CM, Ben Zakour NL, Weinert LA, Conway-Morris A, Cartwright RA, et al. Recent human-to-poultry host jump, adaptation, and pandemic spread of *Staphylococcus aureus*. Proc Natl Acad Sci USA. 2009;106(46):19545–50.

[37] Argudin MA, Mendoza MC, Rodicio MR. Food poisoning and *Staphylococcus aureus* enterotoxins. Toxins (Basel). 2010;2(7):1751–73.

[38] Molla B, Byrne M, Abley M, Mathews J, Jackson CR, Fedorka-Cray P, et al. Epidemiology and genotypic characteristics of methicillin-resistant *Staphylococcus aureus* strains of porcine origin. J Clin Microbiol. 2012;50(11):3687–93.

[39] de Jonge R, Verdier JE, Havelaar AH. Prevalence of meticillin-resistant *Staphylococcus aureus* amongst professional meat handlers in the Netherlands, March-July 2008. Euro Surveill. 2010;15(46):1–5.

[40] Witte W, Cuny C. Emergence and spread of cfr-mediated multiresistance in staphylococci: an interdisciplinary challenge. Future Microbiol. 2011;6(8):925–31.

[41] van Cleef BA, Graveland H, Haenen AP, van de Giessen AW, Heederik D, Wagenaar JA, et al. Persistence of livestock-associated methicillin-resistant *Staphylococcus aureus* in field workers after short-term occupational exposure to pigs and veal calves. J Clin Microbiol. 2011;49(3):1030–3.

[42] Huber H, Giezendanner N, Stephan R, Zweifel C. Genotypes, antibiotic resistance profiles and microarray-based characterization of methicillin-resistant *Staphylococcus aureus* strains isolated from livestock and veterinarians in Switzerland. Zoonoses Public Health. 2011;58(5):343–9.

[43] Fossum Moen AE, Saltyte Benth J, Alm-Kristiansen K, Bukholm G. Exotoxin-encoding gene content in community-acquired and hospital-acquired methicillin-resistant *Staphylococcus aureus*. Clin Microbiol Infect. 2009;15(12):1139–45.

[44] Kadlec K, Ehricht R, Monecke S, Steinacker U, Kaspar H, Mankertz J, et al. Diversity of antimicrobial resistance pheno- and genotypes of methicillin-resistant *Staphylococcus aureus* ST398 from diseased swine. J Antimicrob Chemother. 2009;64(6):1156–64.

[45] Alt K, Fetsch A, Schroeter A, Guerra B, Hammerl JA, Hertwig S, et al. Factors associated with the occurrence of MRSA CC398 in herds of fattening pigs in Germany. BMC Vet Res. 2011;7:69.

[46] Kock R, Harlizius J, Bressan N, Laerberg R, Wieler LH, Witte W, et al. Prevalence and molecular characteristics of methicillin-resistant *Staphylococcus aureus* (MRSA) among pigs on German farms and import of livestock-related MRSA into hospitals. Eur J Clin Microbiol Infect Dis. 2009;28(11):1375–82.

[47] Garcia-Graells C, Antoine J, Larsen J, Catry B, Skov R, Denis O. Livestock veterinarians at high risk of acquiring methicillin-resistant *Staphylococcus aureus* ST398. Epidemiol Infect. 2012;140(3):383–9.

[48] Cuny C, Nathaus R, Layer F, Strommenger B, Altmann D, Witte W. Nasal colonization of humans with methicillin-resistant *Staphylococcus aureus* (MRSA) CC398 with and without exposure to pigs. PLoS One. 2009;4(8):e6800.

[49] IV VDB, BA VANC, Haenen A, Broens EM, PJ VDW, MJ VDB, et al. Methicillin-resistant *Staphylococcus aureus* in people living and working in pig farms. Epidemiol Infect. 2009;137(5):700–8.

[50] Wulf MW, Sorum M, van Nes A, Skov R, Melchers WJ, Klaassen CH, et al. Prevalence of methicillin-resistant *Staphylococcus aureus* among veterinarians: an international study. Clin Microbiol Infect. 2008;14(1):29–34.

[51] Smith TC, Male MJ, Harper AL, Kroeger JS, Tinkler GP, Moritz ED, et al. Methicillin-resistant *Staphylococcus aureus* (MRSA) strain ST398 is present in midwestern U.S. swine and swine workers. PLoS One. 2009;4(1):e4258.

[52] Hanson BM, Dressler AE, Harper AL, Scheibel RP, Wardyn SE, Roberts LK, et al. Prevalence of *Staphylococcus aureus* and methicillin-resistant *Staphylococcus aureus* (MRSA) on retail meat in Iowa. J Infect Public Health. 2011;4(4):169–74.

[53] Uhlemann AC, Porcella SF, Trivedi S, Sullivan SB, Hafer C, Kennedy AD, et al. Identification of a highly transmissible animal-independent *Staphylococcus aureus* ST398 clone with distinct genomic and cell adhesion properties. MBio. 2012;3(2):e00027–12.

[54] Bhat M, Dumortier C, Taylor BS, Miller M, Vasquez G, Yunen J, et al. *Staphylococcus aureus* ST398, New York City and Dominican Republic. Emerg Infect Dis. 2009;15(2):285–7.

[55] Dressler AE, Scheibel RP, Wardyn S, Harper AL, Hanson BM, Kroeger JS, et al. Prevalence, antibiotic resistance and molecular characterisation of *Staphylococcus aureus* in pigs at agricultural fairs in the USA. Vet Rec. 2012;170(19):495.

[56] Voss A, Loeffen F, Bakker J, Klaassen C, Wulf M. Methicillin-resistant *Staphylococcus aureus* in pig farming. Emerg Infect Dis. 2005;11(12):1965–6.

http://dx.doi.org/10.5772/66645

[57] van Duijkeren E, Ikawaty R, Broekhuizen-Stins MJ, Jansen MD, Spalburg EC, de Neeling AJ, et al. Transmission of methicillin-resistant *Staphylococcus aureus* strains between different kinds of pig farms. Vet Microbiol. 2008;126(4):383–9.

[58] Guardabassi L, Stegger M, Skov R. Retrospective detection of methicillin resistant and susceptible *Staphylococcus aureus* ST398 in Danish slaughter pigs. Vet Microbiol. 2007;122(3–4):384–6.

[59] Meemken D, Cuny C, Witte W, Eichler U, Staudt R, Blaha T. Occurrence of MRSA in pigs and in humans involved in pig production—preliminary results of a study in the northwest of Germany. Dtsch Tierarztl Wochenschr. 2008;115(4):132–9.

[60] Khanna T, Friendship R, Dewey C, Weese JS. Methicillin resistant *Staphylococcus aureus* colonization in pigs and pig farmers. Vet Microbiol. 2008;128(3–4):298–303.

[61] European Food Safety Authority. Analysis of the baseline survey on the prevalence of methicillin-resistant *Staphylococcus aureus* (MRSA) in holdings with breeding pigs in the EU, 2008; Part A: MRSA prevalence estimates. EFSA J. 2009;7:1376.

[62] Battisti A, Franco A, Merialdi G, Hasman H, Iurescia M, Lorenzetti R, et al. Heterogeneity among methicillin-resistant *Staphylococcus aureus* from Italian pig finishing holdings. Vet Microbiol. 2010;142(3–4):361–6.

[63] Oppliger A, Moreillon P, Charriere N, Giddey M, Morisset D, Sakwinska O. Antimicrobial resistance of *Staphylococcus aureus* strains acquired by pig farmers from pigs. Appl Environ Microbiol. 2012;78(22):8010–4.

[64] Broens EM, Graat EA, Van der Wolf PJ, Van de Giessen AW, De Jong MC. Prevalence and risk factor analysis of livestock associated MRSA-positive pig herds in The Netherlands. Prev Vet Med. 2011;102(1):41–9.

[65] European Food Safety Authority. Analysis of the baseline survey on the prevalence of methicillin-resistant *Staphylococcus aureus* (MRSA) in holdings with breeding pigs, in the EU, 2008. Part B. Factors associated with MRSA contamination of holdings. EFSA J 2010;8:1597–663.

[66] Crombe F, Willems G, Dispas M, Hallin M, Denis O, Suetens C, et al. Prevalence and antimicrobial susceptibility of methicillin-resistant *Staphylococcus aureus* among pigs in Belgium. Microb Drug Resist. 2012;18(2):125–31.

[67] Broens EM, Graat EA, Van der Wolf PJ, Van de Giessen AW, De Jong MC. Transmission of methicillin resistant *Staphylococcus aureus* among pigs during transportation from farm to abattoir. Vet J. 2011;189(3):302–5.

[68] Espinosa-Gongora C, Broens EM, Moodley A, Nielsen JP, Guardabassi L. Transmission of MRSA CC398 strains between pig farms related by trade of animals. Vet Rec. 2012;170(22):564.

[69] Weese JS, Zwambag A, Rosendal T, Reid-Smith R, Friendship R. Longitudinal investigation of methicillin-resistant *Staphylococcus aureus* in piglets. Zoonoses Public Health. 2011;58(4):238–43.

[70] van Duijkeren E, Jansen MD, Flemming SC, de Neeling H, Wagenaar JA, Schoormans AH, et al. Methicillin-resistant *Staphylococcus aureus* in pigs with exudative epidermitis. Emerg Infect Dis. 2007;13(9):1408–10.

[71] Schwarz S, Kadlec K, Strommenger B. Methicillin-resistant *Staphylococcus aureus* and *Staphylococcus pseudintermedius* detected in the BfT-GermVet monitoring programme 2004–2006 in Germany. J Antimicrob Chemother. 2008;61(2):282–5.

[72] Weese JS, Reid-Smith R, Rousseau J, Avery B. Methicillin-resistant *Staphylococcus aureus* (MRSA) contamination of retail pork. Can Vet J. 2010;51(7):749–52.

[73] Osadebe LU, Hanson B, Smith TC, Heimer R. Prevalence and characteristics of *Staphylococcus aureus* in Connecticut swine and swine farmers. Zoonoses Public Health. 2013;60(3):234–43.

[74] Gordoncillo MJ, Abdujamilova N, Perri M, Donabedian S, Zervos M, Bartlett P. Detection of methicillin-resistant *Staphylococcus aureus* (MRSA) in backyard pigs and their owners, Michigan, USA. Zoonoses Public Health. 2012;59(3):212–6.

[75] Frana TS, Beahm AR, Hanson BM, Kinyon JM, Layman LL, Karriker LA, et al. Isolation and characterization of methicillin-resistant *Staphylococcus aureus* from pork farms and visiting veterinary students. PLoS One. 2013;8(1):e53738.

[76] Smith TC, Gebreyes WA, Abley MJ, Harper AL, Forshey BM, Male MJ, et al. Methicillin-resistant *Staphylococcus aureus* in pigs and farm workers on conventional and antibiotic-free swine farms in the USA. PLoS One. 2013;8(5):e63704.

[77] Rinsky JL, Nadimpalli M, Wing S, Hall D, Baron D, Price LB, et al. Livestock-associated methicillin and multidrug resistant *Staphylococcus aureus* is present among industrial, not antibiotic-free livestock operation workers in North Carolina. PLoS One. 2013;8(7):e67641.

[78] Neyra RC, Frisancho JA, Rinsky JL, Resnick C, Carroll KC, Rule AM, et al. Multidrug-resistant and methicillin-resistant *Staphylococcus aureus* (MRSA) in hog slaughter and processing plant workers and their community in North Carolina (USA). Environ Health Perspect. 2014;122(5):471–7.

[79] Nadimpalli M, Rinsky JL, Wing S, Hall D, Stewart J, Larsen J, et al. Persistence of livestock-associated antibiotic-resistant *Staphylococcus aureus* among industrial hog operation workers in North Carolina over 14 days. Occup Environ Med. 2015;72(2):90–9.

[80] Cui S, Li J, Hu C, Jin S, Li F, Guo Y, et al. Isolation and characterization of methicillin-resistant *Staphylococcus aureus* from swine and workers in China. J Antimicrob Chemother. 2009;64(4):680–3.

[81] Ho PL, Chow KH, Lai EL, Law PY, Chan PY, Ho AY, et al. Clonality and antimicrobial susceptibility of *Staphylococcus aureus* and methicillin-resistant *S. aureus* isolates from food animals and other animals. J Clin Microbiol. 2012;50(11):3735–7.

[82] Sergio DM, Koh TH, Hsu LY, Ogden BE, Goh AL, Chow PK. Investigation of meticillin-resistant *Staphylococcus aureus* in pigs used for research. J Med Microbiol. 2007;56(Pt 8):1107–9.

[83] Hsu LY, Koh TH, Singh K, Kang ML, Kurup A, Tan BH. Dissemination of multisusceptible methicillin-resistant *Staphylococcus aureus* in Singapore. J Clin Microbiol. 2005;43(6):2923–5.

[84] Tsai HY, Liao CH, Cheng A, Liu CY, Huang YT, Teng LJ, et al. Isolation of meticillin-resistant *Staphylococcus aureus* sequence type 9 in pigs in Taiwan. Int J Antimicrob Agents. 2012;39(5):449–51.

[85] Lim SK, Nam HM, Jang GC, Lee HS, Jung SC, Kwak HS. The first detection of methicillin-resistant *Staphylococcus aureus* ST398 in pigs in Korea. Vet Microbiol. 2012;155(1):88–92.

[86] Devriese LA, Van Damme LR, Fameree L. Methicillin (cloxacillin)-resistant *Staphylococcus aureus* strains isolated from bovine mastitis cases. Zentralbl Veterinarmed B. 1972;19(7):598–605.

[87] Vanderhaeghen W, Cerpentier T, Adriaensen C, Vicca J, Hermans K, Butaye P. Methicillin-resistant *Staphylococcus aureus* (MRSA) ST398 associated with clinical and subclinical mastitis in Belgian cows. Vet Microbiol. 2010;144(1–2):166–71.

[88] Devriese LA, Hommez J. Epidemiology of methicillin-resistant *Staphylococcus aureus* in dairy herds. Res Vet Sci. 1975;19(1):23–7.

[89] Lee JH. Methicillin (Oxacillin)-resistant *Staphylococcus aureus* strains isolated from major food animals and their potential transmission to humans. Appl Environ Microbiol. 2003;69(11):6489–94.

[90] Moon JS, Lee AR, Kang HM, Lee ES, Kim MN, Paik YH, et al. Phenotypic and genetic antibiogram of methicillin-resistant staphylococci isolated from bovine mastitis in Korea. J Dairy Sci. 2007;90(3):1176–85.

[91] Moon DC, Tamang MD, Nam HM, Jeong JH, Jang GC, Jung SC, et al. Identification of livestock-associated methicillin-resistant *Staphylococcus aureus* isolates in Korea and molecular comparison between isolates from animal carcasses and slaughterhouse workers. Foodborne Pathog Dis. 2015;12(4):327–34.

[92] Juhasz-Kaszanyitzky E, Janosi S, Somogyi P, Dan A, van der Graaf-van Bloois L, van Duijkeren E, et al. MRSA transmission between cows and humans. Emerg Infect Dis. 2007;13(4):630–2.

[93] Weese JS, Archambault M, Willey BM, Hearn P, Kreiswirth BN, Said-Salim B, et al. Methicillin-resistant *Staphylococcus aureus* in horses and horse personnel, 2000–2002. Emerg Infect Dis. 2005;11(3):430–5.

[94] Umoh VJ, Adesiyun AA, Gomwalk NE. Antibiogram of staphylococcal strains isolated from milk and milk-products. Zentralbl Veterinarmed B. 1990;37(9):701–6.

[95] Costa EO, Benites NR, Guerra JL, Melville PA. Antimicrobial susceptibility of *Staphylococcus* spp. isolated from mammary parenchymas of slaughtered dairy cows. J Vet Med B Infect Dis Vet Public Health. 2000;47(2):99–103.

[96] Erskine RJ, Walker RD, Bolin CA, Bartlett PC, White DG. Trends in antibacterial susceptibility of mastitis pathogens during a seven-year period. J Dairy Sci. 2002;85(5):1111–8.

[97] Fessler A, Scott C, Kadlec K, Ehricht R, Monecke S, Schwarz S. Characterization of methicillin-resistant *Staphylococcus aureus* ST398 from cases of bovine mastitis. J Antimicrob Chemother. 2010;65(4):619–25.

[98] Bos ME, Graveland H, Portengen L, Wagenaar JA, Heederik DJ. Livestock-associated MRSA prevalence in veal calf production is associated with farm hygiene, use of antimicrobials, and age of the calves. Prev Vet Med. 2012;105(1–2):155–9.

[99] Eurosurveillance editorial team. European Union Summary Report on antimicrobial resistance in zoonotic and indicator bacteria from humans, animals and food 2012 published. Euro Surveill. 2014;19(12):20748.

[100] Graveland H, Wagenaar JA, Bergs K, Heesterbeek H, Heederik D. Persistence of livestock associated MRSA CC398 in humans is dependent on intensity of animal contact. PLoS One. 2011;6(2):e16830.

[101] Holmes MA, Zadoks RN. Methicillin resistant *S. aureus* in human and bovine mastitis. J Mammary Gland Biol Neoplasia. 2011;16(4):373–82.

[102] Spohr M, Rau J, Friedrich A, Klittich G, Fetsch A, Guerra B, et al. Methicillin-resistant *Staphylococcus aureus* (MRSA) in three dairy herds in southwest Germany. Zoonoses Public Health. 2011;58(4):252–61.

[103] Monecke S, Kuhnert P, Hotzel H, Slickers P, Ehricht R. Microarray based study on virulence-associated genes and resistance determinants of *Staphylococcus aureus* isolates from cattle. Vet Microbiol. 2007;125(1–2):128–40.

[104] Hata E, Katsuda K, Kobayashi H, Uchida I, Tanaka K, Eguchi M. Genetic variation among *Staphylococcus aureus* strains from bovine milk and their relevance to methicillin-resistant isolates from humans. J Clin Microbiol. 2010;48(6):2130–9.

[105] van Rijen MM, Bosch T, Heck ME, Kluytmans JA. Meticillin-resistant *Staphylococcus aureus*_epidemiology and transmission in a Dutch hospital. J Hosp Infect. 2009;72(4):299–306.

[106] van Rijen MM, Van Keulen PH, Kluytmans JA. Increase in a Dutch hospital of methicillin-resistant *Staphylococcus aureus* related to animal farming. Clin Infect Dis. 2008;46(2):261–3.

[107] Vandenbroucke-Grauls CM, Beaujean DJ. Methicillin-resistant *Staphylococcus aureus* in pig breeders and cattle breeders. Ned Tijdschr Geneeskd. 2006;150(31):1710–2.

[108] Nam HM, Lee AL, Jung SC, Kim MN, Jang GC, Wee SH, et al. Antimicrobial susceptibility of *Staphylococcus aureus* and characterization of methicillin-resistant *Staphylococcus aureus* isolated from bovine mastitis in Korea. Foodborne Pathog Dis. 2011;8(2):231–8.

[109] Kwon NH, Park KT, Moon JS, Jung WK, Kim SH, Kim JM, et al. Staphylococcal cassette chromosome mec (SCCmec) characterization and molecular analysis for methicillin-resistant *Staphylococcus aureus* and novel SCCmec subtype IVg isolated from bovine milk in Korea. J Antimicrob Chemother. 2005;56(4):624–32.

[110] Feltrin F, Alba P, Kraushaar B, Ianzano A, Argudin MA, Di Matteo P, et al. A livestock-associated, multidrug-resistant, methicillin-resistant *Staphylococcus aureus* clonal complex 97 lineage spreading in dairy cattle and pigs in Italy. Appl Environ Microbiol. 2016;82(3):816–21.

[111] Garcia-Alvarez L, Holden MT, Lindsay H, Webb CR, Brown DF, Curran MD, et al. Meticillin-resistant *Staphylococcus aureus* with a novel mecA homologue in human and bovine populations in the UK and Denmark: a descriptive study. Lancet Infect Dis. 2011;11(8):595–603.

[112] Petersen A, Stegger M, Heltberg O, Christensen J, Zeuthen A, Knudsen LK, et al. Epidemiology of methicillin-resistant *Staphylococcus aureus* carrying the novel mecC gene in Denmark corroborates a zoonotic reservoir with transmission to humans. Clin Microbiol Infect. 2013;19(1):E16–22.

[113] Weese JS, Hannon SJ, Booker CW, Gow S, Avery BP, Reid-Smith RJ. The prevalence of methicillin-resistant *Staphylococcus aureus* colonization in feedlot cattle. Zoonoses Public Health. 2012;59(2):144–7.

[114] Golding GR, Levett PN, McDonald RR, Wylie J, Mulvey MR, editors. Molecular characterization of the first clinical infections of "livestock associated" ST398 methicillin-resistant *Staphylococcus aureus* in Canada. American Society for Microbiology (ASM)-ESCMID Conference on Methicillin-Resistant Staphylococci in Animals; London.

[115] Lee JH. Occurrence of methicillin-resistant *Staphylococcus aureus* strains from cattle and chicken, and analyses of their mecA, mecR1 and mecI genes. Vet Microbiol. 2006;114(1–2):155–9.

[116] Alves PD, McCulloch JA, Even S, Le Marechal C, Thierry A, Grosset N, et al. Molecular characterisation of *Staphylococcus aureus* strains isolated from small and large ruminants reveals a host rather than tissue specificity. Vet Microbiol. 2009;137(1–2):190–5.

[117] Haenni M, Saras E, Chatre P, Medaille C, Bes M, Madec JY, et al. A USA300 variant and other human-related methicillin-resistant *Staphylococcus aureus* strains infecting cats and dogs in France. J Antimicrob Chemother. 2012;67(2):326–9.

[118] Lin Y, Barker E, Kislow J, Kaldhone P, Stemper ME, Pantrangi M, et al. Evidence of multiple virulence subtypes in nosocomial and community-associated MRSA genotypes in

companion animals from the upper midwestern and northeastern United States. Clin Med Res. 2011;9(1):7–16.

[119] Alba P, Feltrin F, Cordaro G, Porrero MC, Kraushaar B, Argudin MA, et al. Livestock-associated methicillin resistant and methicillin susceptible *Staphylococcus aureus* sequence type (CC)1 in European farmed animals: high genetic relatedness of isolates from Italian cattle herds and humans. PLoS One. 2015;10(8):e0137143.

[120] Heikinheimo A, Johler S, Karvonen L, Julmi J, Fredriksson-Ahomaa M, Stephan R. New dominant spa type t2741 in livestock-associated MRSA (CC398-MRSA-V) in Finnish fattening pigs at slaughter. Antimicrob Resist Infect Control. 2016;5:6.

[121] Peeters LE, Argudin MA, Azadikhah S, Butaye P. Antimicrobial resistance and population structure of *Staphylococcus aureus* recovered from pigs farms. Vet Microbiol. 2015;180(1–2):151–6.

[122] Kitai S, Shimizu A, Kawano J, Sato E, Nakano C, Kitagawa H, et al. Prevalence and characterization of *Staphylococcus aureus* and enterotoxigenic *Staphylococcus aureus* in retail raw chicken meat throughout Japan. J Vet Med Sci. 2005;67(3):269–74.

[123] Spoor LE, McAdam PR, Weinert LA, Rambaut A, Hasman H, Aarestrup FM, et al. Livestock origin for a human pandemic clone of community-associated methicillin-resistant *Staphylococcus aureus*. MBio. 2013;4(4):e00356–13.

[124] Pletinckx LJ, Verhegghe M, Dewulf J, Crombe F, De Bleecker Y, Rasschaert G, et al. Screening of poultry-pig farms for methicillin-resistant *Staphylococcus aureus*: sampling methodology and within herd prevalence in broiler flocks and pigs. Infect Genet Evol. 2011;11(8):2133–7.

[125] Persoons D, Van Hoorebeke S, Hermans K, Butaye P, de Kruif A, Haesebrouck F, et al. Methicillin-resistant *Staphylococcus aureus* in poultry. Emerg Infect Dis. 2009;15(3):452–3.

[126] Vossenkuhl B, Brandt J, Fetsch A, Kasbohrer A, Kraushaar B, Alt K, et al. Comparison of spa types, SCCmec types and antimicrobial resistance profiles of MRSA isolated from turkeys at farm, slaughter and from retail meat indicates transmission along the production chain. PLoS One. 2014;9(5):e96308.

[127] Pu S, Han F, Ge B. Isolation and characterization of methicillin-resistant *Staphylococcus aureus* strains from Louisiana retail meats. Appl Environ Microbiol. 2009;75(1):265–7.

[128] Lin J, Yeh KS, Liu HT, Lin JH. *Staphylococcus aureus* isolated from pork and chicken carcasses in Taiwan: prevalence and antimicrobial susceptibility. J Food Prot. 2009;72(3):608–11.

[129] Hasman H, Moodley A, Guardabassi L, Stegger M, Skov RL, Aarestrup FM. Spa type distribution in *Staphylococcus aureus* originating from pigs, cattle and poultry. Vet Microbiol. 2010;141(3–4):326–31.

[130] Cavaco LM, Hasman H, Aarestrup FM. Zinc resistance of *Staphylococcus aureus* of animal origin is strongly associated with methicillin resistance. Vet Microbiol. 2011;150(3–4):344–8.

[131] Smyth DS, Feil EJ, Meaney WJ, Hartigan PJ, Tollersrud T, Fitzgerald JR, et al. Molecular genetic typing reveals further insights into the diversity of animal-associated *Staphylococcus aureus*. J Med Microbiol. 2009;58(Pt 10):1343–53.

[132] Mulders MN, Haenen AP, Geenen PL, Vesseur PC, Poldervaart ES, Bosch T, et al. Prevalence of livestock-associated MRSA in broiler flocks and risk factors for slaughterhouse personnel in The Netherlands. Epidemiol Infect. 2010;138(5):743–55.

[133] Pletinckx LJ, Verhegghe M, Crombe F, Dewulf J, De Bleecker Y, Rasschaert G, et al. Evidence of possible methicillin-resistant *Staphylococcus aureus* ST398 spread between pigs and other animals and people residing on the same farm. Prev Vet Med. 2013;109(3–4):293–303.

[134] Wendlandt S, Li J, Ho J, Porta MA, Fessler AT, Wang Y, et al. Enterococcal multiresistance gene cluster in methicillin-resistant *Staphylococcus aureus* from various origins and geographical locations. J Antimicrob Chemother. 2014;69(9):2573–5.

[135] Kwon NH, Park KT, Jung WK, Youn HY, Lee Y, Kim SH, et al. Characteristics of methicillin resistant *Staphylococcus aureus* isolated from chicken meat and hospitalized dogs in Korea and their epidemiological relatedness. Vet Microbiol. 2006;117(2–4):304–12.

[136] Ekkelenkamp MB, Sekkat M, Carpaij N, Troelstra A, Bonten MJ. Endocarditis due to meticillin-resistant *Staphylococcus aureus* originating from pigs. Ned Tijdschr Geneeskd. 2006;150(44):2442–7.

[137] de Boer E, Zwartkruis-Nahuis JT, Wit B, Huijsdens XW, de Neeling AJ, Bosch T, et al. Prevalence of methicillin-resistant *Staphylococcus aureus* in meat. Int J Food Microbiol. 2009;134(1–2):52–6.

[138] van Loo IH, Diederen BM, Savelkoul PH, Woudenberg JH, Roosendaal R, van Belkum A, et al. Methicillin-resistant *Staphylococcus aureus* in meat products, the Netherlands. Emerg Infect Dis. 2007;13(11):1753–5.

[139] Schraft H, Kleinlein N, Untermann F. Contamination of pig hindquarters with *Staphylococcus aureus*. Int J Food Microbiol. 1992;15(1–2):191–4.

[140] Hendriksen RS, Mevius DJ, Schroeter A, Teale C, Meunier D, Butaye P, et al. Prevalence of antimicrobial resistance among bacterial pathogens isolated from cattle in different European countries: 2002–2004. Acta Vet Scand. 2008;50:28.

[141] Tenhagen BA, Vossenkuhl B, Kasbohrer A, Alt K, Kraushaar B, Guerra B, et al. Methicillin-resistant *Staphylococcus aureus* in cattle food chains—prevalence, diversity, and antimicrobial resistance in Germany. J Anim Sci. 2014;92(6):2741–51.

[142] Foster AP. Staphylococcal skin disease in livestock. Vet Dermatol. 2012;23(4):342–51, e63.

[143] Atyah MA, Zamri-Saad M, Siti-Zahrah A. First report of methicillin-resistant *Staphylococcus aureus* from cage-cultured tilapia (*Oreochromis niloticus*). Vet Microbiol. 2010;144(3–4):502–4.

[144] Rhee CH, Woo GJ. Emergence and characterization of foodborne methicillin-resistant *Staphylococcus aureus* in Korea. J Food Prot. 2010;73(12):2285–90.

[145] Hammad AM, Watanabe W, Fujii T, Shimamoto T. Occurrence and characteristics of methicillin-resistant and -susceptible *Staphylococcus aureus* and methicillin-resistant coagulase-negative staphylococci from Japanese retail ready-to-eat raw fish. Int J Food Microbiol. 2012;156(3):286–9.

[146] Sergelidis D, Abrahim A, Papadopoulos T, Soultos N, Martziou E, Koulourida V, et al. Isolation of methicillin-resistant *Staphylococcus* spp. from ready-to-eat fish products. Lett Appl Microbiol. 2014;59(5):500–6.

[147] Sugimoto K, Frei R, Graber P. Methicillin-resistant *Staphylococcus aureus* foot infection after fish pedicure. Infection. 2013;41(5):1013–5.

[148] Devriese LA. A simplified system for biotyping *Staphylococcus aureus* strains isolated from animal species. J Appl Bacteriol. 1984;56(2):215–20.

[149] Bhargava K, Wang X, Donabedian S, Zervos M, de Rocha L, Zhang Y. Methicillin-resistant *Staphylococcus aureus* in retail meat, Detroit, Michigan, USA. Emerg Infect Dis. 2011;17(6):1135–7.

[150] Kelman A, Soong YA, Dupuy N, Shafer D, Richbourg W, Johnson K, et al. Antimicrobial susceptibility of *Staphylococcus aureus* from retail ground meats. J Food Prot. 2011;74(10):1625–9.

[151] Carr MA, Thompson LD, Miller MF, Ramsey CB, Kaster CS. Chilling and trimming effects on the microbial populations of pork carcasses. J Food Prot. 1998;61(4):487–9.

[152] Beneke B, Klees S, Stuhrenberg B, Fetsch A, Kraushaar B, Tenhagen BA. Prevalence of methicillin-resistant *Staphylococcus aureus* in a fresh meat pork production chain. J Food Prot. 2011;74(1):126–9.

[153] Borch E, Nesbakken T, Christensen H. Hazard identification in swine slaughter with respect to foodborne bacteria. Int J Food Microbiol. 1996;30(1–2):9–25.

[154] Spescha C, Stephan R, Zweifel C. Microbiological contamination of pig carcasses at different stages of slaughter in two European Union-approved abattoirs. J Food Prot. 2006;69(11):2568–75.

[155] Bolton DJ, Pearce RA, Sheridan JJ, Blair IS, McDowell DA, Harrington D. Washing and chilling as critical control points in pork slaughter hazard analysis and critical control point (HACCP) systems. J Appl Microbiol. 2002;92(5):893–902.

[156] Vossenkuhl B, Sharp H, Brandt J, Fetsch A, Käsbohrer A, Tenhagen BA. Modeling the transmission of livestock associated methicillin-resistant *Staphylococcus aureus* along the pig slaughter line. Food Control. 2014;39:17–24.

[157] Dockerty TR, Ockerman HW, Cahill VR, Kunkle LE, Weiser HH. Microbial level of pork skin as affected by the dressing process. J Anim Sci 1970;30:884–90.

[158] Pearce RA, Bolton DJ, Sheridan JJ, McDowell DA, Blair IS, Harrington D. Studies to determine the critical control points in pork slaughter hazard analysis and critical control point systems. Int J Food Microbiol. 2004;90(3):331–9.

[159] Yu SL, Bolton D, Laubach C, Kline P, Oser A, Palumbo SA. Effect of dehairing operations on microbiological quality of swine carcasses. J Food Prot. 1999;62(12):1478–81.

[160] Saide-Albornoz JJ, Lynn Knipe C, Murano EA, Beran GW. Contamination of pork carcasses during slaughter, fabrication, and chilled storage. J Food Prot 1995;58:993–7.

[161] Rivas T, Vizcaino JA, Herrera FJ. Microbial contamination of carcasses and equipment from an Iberian pig slaughterhouse. J Food Prot. 2000;63(12):1670–5.

[162] Brown T, James SJ. Process design data for pork chilling. Int J Refrig. 1992;15:281–9.

[163] James S. The chill chain "from carcass to consumer". Meat Sci. 1996;43S1:203–16.

[164] Graveland H, Duim B, van Duijkeren E, Heederik D, Wagenaar JA. Livestock-associated methicillin-resistant *Staphylococcus aureus* in animals and humans. Int J Med Microbiol. 2011;301(8):630–4.

[165] Lewis HC, Molbak K, Reese C, Aarestrup FM, Selchau M, Sorum M, et al. Pigs as source of methicillin-resistant *Staphylococcus aureus* CC398 infections in humans, Denmark. Emerg Infect Dis. 2008;14(9):1383–9.

[166] Cefai C, Ashurst S, Owens C. Human carriage of methicillin-resistant *Staphylococcus aureus* linked with pet dog. Lancet. 1994;344(8921):539–40.

[167] Price LB, Stegger M, Hasman H, Aziz M, Larsen J, Andersen PS, et al. *Staphylococcus aureus* CC398: host adaptation and emergence of methicillin resistance in livestock. MBio. 2012;3(1):e00305–11.

[168] Rooijakkers SH, Ruyken M, Roos A, Daha MR, Presanis JS, Sim RB, et al. Immune evasion by a staphylococcal complement inhibitor that acts on C3 convertases. Nat Immunol. 2005;6(9):920–7.

[169] Stegger M, Lindsay JA, Moodley A, Skov R, Broens EM, Guardabassi L. Rapid PCR detection of *Staphylococcus aureus* clonal complex 398 by targeting the restriction-modification system carrying sau1-hsdS1. J Clin Microbiol. 2011;49(2):732–4.

[170] Waldron DE, Lindsay JA. Sau1: a novel lineage-specific type I restriction-modification system that blocks horizontal gene transfer into *Staphylococcus aureus* and between *S. aureus* isolates of different lineages. J Bacteriol. 2006;188(15):5578–85.

[171] Waters AE, Contente-Cuomo T, Buchhagen J, Liu CM, Watson L, Pearce K, et al. Multidrug-resistant *Staphylococcus aureus* in US Meat and Poultry. Clin Infect Dis. 2011;52(10):1227–30.

[172] Wieler LH, Ewers C, Guenther S, Walther B, Lubke-Becker A. Methicillin-resistant staphylococci (MRS) and extended-spectrum beta-lactamases (ESBL)-producing Enterobacteriaceae in companion animals: nosocomial infections as one reason for the rising prevalence of these potential zoonotic pathogens in clinical samples. Int J Med Microbiol. 2011;301(8):635–41.

[173] Takeuchi S, Matsunaga K, Inubushi S, Higuchi H, Imaizumi K, Kaidoh T. Structural gene and strain specificity of a novel cysteine protease produced by *Staphylococcus aureus* isolated from a diseased chicken. Vet Microbiol. 2002;89(2–3):201–10.

[174] Ward MJ, Gibbons CL, McAdam PR, van Bunnik BA, Girvan EK, Edwards GF, et al. Time-scaled evolutionary analysis of the transmission and antibiotic resistance dynamics of *Staphylococcus aureus* clonal complex 398. Appl Environ Microbiol. 2014;80(23):7275–82.

[175] Jansen MD, Box AT, Fluit AC. SCCmec typing in methicillin-resistant *Staphylococcus aureus* strains of animal origin. Emerg Infect Dis. 2009;15(1):136–7; author reply-7.

[176] Zhang Y, Agidi S, LeJeune JT. Diversity of staphylococcal cassette chromosome in coagulase-negative staphylococci from animal sources. J Appl Microbiol. 2009;107(4):1375–83.

[177] Guardabassi L, O'Donoghue M, Moodley A, Ho J, Boost M. Novel lineage of methicillin-resistant *Staphylococcus aureus*, Hong Kong. Emerg Infect Dis. 2009;15(12):1998–2000.

[178] Fitzgerald JR, Sturdevant DE, Mackie SM, Gill SR, Musser JM. Evolutionary genomics of *Staphylococcus aureus*: insights into the origin of methicillin-resistant strains and the toxic shock syndrome epidemic. Proc Natl Acad Sci USA. 2001;98(15):8821–6.

[179] Oliveira DC, Tomasz A, de Lencastre H. Secrets of success of a human pathogen: molecular evolution of pandemic clones of meticillin-resistant *Staphylococcus aureus*. Lancet Infect Dis. 2002;2(3):180–9.

[180] Nubel U, Roumagnac P, Feldkamp M, Song JH, Ko KS, Huang YC, et al. Frequent emergence and limited geographic dispersal of methicillin-resistant *Staphylococcus aureus*. Proc Natl Acad Sci USA. 2008;105(37):14130–5.

[181] Rijnders MI, Deurenberg RH, Boumans ML, Hoogkamp-Korstanje JA, Beisser PS, Antibiotic Resistance Surveillance Group, et al. Population structure of *Staphylococcus aureus* strains isolated from intensive care unit patients in the Netherlands over an 11-year period (1996 to 2006). J Clin Microbiol. 2009;47(12):4090–5.

[182] Varshney AK, Mediavilla JR, Robiou N, Guh A, Wang X, Gialanella P, et al. Diverse enterotoxin gene profiles among clonal complexes of *Staphylococcus aureus* isolates from the Bronx, New York. Appl Environ Microbiol. 2009;75(21):6839–49.

[183] Walther B, Monecke S, Ruscher C, Friedrich AW, Ehricht R, Slickers P, et al. Comparative molecular analysis substantiates zoonotic potential of equine methicillin-resistant *Staphylococcus aureus*. J Clin Microbiol. 2009;47(3):704–10.

[184] McCarthy AJ, Witney AA, Gould KA, Moodley A, Guardabassi L, Voss A, et al. The distribution of mobile genetic elements (MGEs) in MRSA CC398 is associated with both host and country. Genome Biol Evol. 2011;3:1164–74.

[185] van Belkum A, Melles DC, Peeters JK, van Leeuwen WB, van Duijkeren E, Huijsdens XW, et al. Methicillin-resistant and -susceptible *Staphylococcus aureus* sequence type 398 in pigs and humans. Emerg Infect Dis. 2008;14(3):479–83.

[186] Welinder-Olsson C, Floren-Johansson K, Larsson L, Oberg S, Karlsson L, Ahren C. Infection with Panton-Valentine leukocidin-positive methicillin-resistant *Staphylococcus aureus* t034. Emerg Infect Dis. 2008;14(8):1271–2.

[187] Yu F, Chen Z, Liu C, Zhang X, Lin X, Chi S, et al. Prevalence of *Staphylococcus aureus* carrying Panton-Valentine leukocidin genes among isolates from hospitalised patients in China. Clin Microbiol Infect. 2008;14(4):381–4.

[188] Ola Ojo M. Bacteriophage types and antibiotic sensitivity of *Staphylococcus aureus* isolated from swabs of the noses and skins of dogs. Vet Rec. 1972;91(6):152–3.

[189] Wang N, Neilan AM, Klompas M. *Staphylococcus intermedius* infections: case report and literature review. Infect Dis Rep. 2013;5(1):e3.

[190] Nagase N, Sasaki A, Yamashita K, Shimizu A, Wakita Y, Kitai S, et al. Isolation and species distribution of staphylococci from animal and human skin. J Vet Med Sci. 2002;64(3):245–50.

[191] Guardabassi L, Schwarz S, Lloyd DH. Pet animals as reservoirs of antimicrobial-resistant bacteria. J Antimicrob Chemother. 2004;54(2):321–32.

[192] Patel A, Lloyd DH, Howell SA, Noble WC. Investigation into the potential pathogenicity of *Staphylococcus felis* in a cat. Vet Rec. 2002;150(21):668–9.

[193] Lilenbaum W, Nunes EL, Azeredo MA. Prevalence and antimicrobial susceptibility of staphylococci isolated from the skin surface of clinically normal cats. Lett Appl Microbiol. 1998;27(4):224–8.

[194] Couto N, Pomba C, Moodley A, Guardabassi L. Prevalence of meticillin-resistant staphylococci among dogs and cats at a veterinary teaching hospital in Portugal. Vet Rec. 2011;169(3):72.

[195] Loeffler A, Pfeiffer DU, Lindsay JA, Soares Magalhaes RJ, Lloyd DH. Prevalence of and risk factors for MRSA carriage in companion animals: a survey of dogs, cats and horses. Epidemiol Infect. 2011;139(7):1019–28.

[196] Loeffler A, Boag AK, Sung J, Lindsay JA, Guardabassi L, Dalsgaard A, et al. Prevalence of methicillin-resistant *Staphylococcus aureus* among staff and pets in a small animal referral hospital in the UK. J Antimicrob Chemother. 2005;56(4):692–7.

[197] Weese JS, Faires M, Rousseau J, Bersenas AM, Mathews KA. Cluster of methicillin-resistant *Staphylococcus aureus* colonization in a small animal intensive care unit. J Am Vet Med Assoc. 2007;231(9):1361–4.

[198] Weese JS, van Duijkeren E. Methicillin-resistant *Staphylococcus aureus* and *Staphylococcus pseudintermedius* in veterinary medicine. Vet Microbiol. 2010;140(3–4):418–29.

[199] Gortel K, Campbell KL, Kakoma I, Whittem T, Schaeffer DJ, Weisiger RM. Methicillin resistance among staphylococci isolated from dogs. Am J Vet Res. 1999;60(12):1526–30.

[200] Griffeth GC, Morris DO, Abraham JL, Shofer FS, Rankin SC. Screening for skin carriage of methicillin-resistant coagulase-positive staphylococci and *Staphylococcus schleiferi* in dogs with healthy and inflamed skin. Vet Dermatol. 2008;19(3):142–9.

[201] Abraham JL, Morris DO, Griffeth GC, Shofer FS, Rankin SC. Surveillance of healthy cats and cats with inflammatory skin disease for colonization of the skin by methicillin-resistant coagulase-positive staphylococci and *Staphylococcus schleiferi* ssp. schleiferi. Vet Dermatol. 2007;18(4):252–9.

[202] Soares Magalhaes RJ, Loeffler A, Lindsay J, Rich M, Roberts L, Smith H, et al. Risk factors for methicillin-resistant *Staphylococcus aureus* (MRSA) infection in dogs and cats: a case-control study. Vet Res. 2010;41(5):55.

[203] Rich M, Roberts L. Methicillin-resistant *Staphylococcus aureus* isolates from companion animals. Vet Rec. 2004;154(10):310.

[204] Quitoco IM, Ramundo MS, Silva-Carvalho MC, Souza RR, Beltrame CO, de Oliveira TF, et al. First report in South America of companion animal colonization by the USA1100 clone of community-acquired meticillin-resistant *Staphylococcus aureus* (ST30) and by the European clone of methicillin-resistant *Staphylococcus pseudintermedius* (ST71). BMC Res Notes. 2013;6:336.

[205] Chanchaithong P, Perreten V, Schwendener S, Tribuddharat C, Chongthaleong A, Niyomtham W, et al. Strain typing and antimicrobial susceptibility of methicillin-resistant coagulase-positive staphylococcal species in dogs and people associated with dogs in Thailand. J Appl Microbiol. 2014;117(2):572–86.

[206] Couto N, Belas A, Kadlec K, Schwarz S, Pomba C. Clonal diversity, virulence patterns and antimicrobial and biocide susceptibility among human, animal and environmental MRSA in Portugal. J Antimicrob Chemother. 2015;70(9):2483–7.

[207] Weiss S, Kadlec K, Fessler AT, Schwarz S. Identification and characterization of methicillin-resistant *Staphylococcus aureus, Staphylococcus epidermidis, Staphylococcus haemolyticus* and *Staphylococcus pettenkoferi* from a small animal clinic. Vet Microbiol. 2013;167(3–4):680–5.

[208] Loncaric I, Kunzel F, Licka T, Simhofer H, Spergser J, Rosengarten R. Identification and characterization of methicillin-resistant *Staphylococcus aureus* (MRSA) from Austrian companion animals and horses. Vet Microbiol. 2014;168(2–4):381–7.

[209] Scott GM, Thomson R, Malone-Lee J, Ridgway GL. Cross-infection between animals and man: possible feline transmission of *Staphylococcus aureus* infection in humans? J Hosp Infect. 1988;12(1):29–34.

[210] Strommenger B, Kehrenberg C, Kettlitz C, Cuny C, Verspohl J, Witte W, et al. Molecular characterization of methicillin-resistant *Staphylococcus aureus* strains from pet animals and their relationship to human isolates. J Antimicrob Chemother. 2006;57(3):461–5.

[211] Weese JS, Dick H, Willey BM, McGeer A, Kreiswirth BN, Innis B, et al. Suspected transmission of methicillin-resistant *Staphylococcus aureus* between domestic pets and humans in veterinary clinics and in the household. Vet Microbiol. 2006;115(1–3):148–55.

[212] Malik S, Coombs GW, O'Brien FG, Peng H, Barton MD. Molecular typing of methicillin-resistant staphylococci isolated from cats and dogs. J Antimicrob Chemother. 2006;58(2):428–31.

[213] Grinberg A, Kingsbury DD, Gibson IR, Kirby BM, Mack HJ, Morrison D. Clinically overt infections with methicillin-resistant *Staphylococcus aureus* in animals in New Zealand: a pilot study. N Z Vet J. 2008;56(5):237–42.

[214] Baptiste KE, Williams K, Willams NJ, Wattret A, Clegg PD, Dawson S, et al. Methicillin-resistant staphylococci in companion animals. Emerg Infect Dis. 2005;11(12):1942–4.

[215] Monaco M, Pedroni P, Sanchini A, Bonomini A, Indelicato A, Pantosti A. Livestock-associated methicillin-resistant *Staphylococcus aureus* responsible for human colonization and infection in an area of Italy with high density of pig farming. BMC Infect Dis. 2013;13:258.

[216] Morris DO, Mauldin EA, O'Shea K, Shofer FS, Rankin SC. Clinical, microbiological, and molecular characterization of methicillin-resistant *Staphylococcus aureus* infections of cats. Am J Vet Res. 2006;67(8):1421–5.

[217] Catry B, Van Duijkeren E, Pomba MC, Greko C, Moreno MA, Pyorala S, et al. Reflection paper on MRSA in food-producing and companion animals: epidemiology and control options for human and animal health. Epidemiol Infect. 2010;138(5):626–44.

[218] Maddox TW, Clegg PD, Williams NJ, Pinchbeck GL. Antimicrobial resistance in bacteria from horses: epidemiology of antimicrobial resistance. Equine Vet J. 2015;47(6):756–65.

[219] Tyner GA, Nolen-Walston RD, Hall T, Palmero JP, Couetil L, Javsicas L, et al. A multicenter retrospective study of 151 renal biopsies in horses. J Vet Intern Med. 2011;25(3):532–9.

[220] Weese JS, Rousseau J, Willey BM, Archambault M, McGeer A, Low DE. Methicillin-resistant *Staphylococcus aureus* in horses at a veterinary teaching hospital: frequency, characterization, and association with clinical disease. J Vet Intern Med. 2006;20(1):182–6.

[221] Haenni M, Chatre P, Dupieux C, Metayer V, Maillard K, Bes M, et al. mecC-positive MRSA in horses. J Antimicrob Chemother. 2015;70(12):3401–2.

[222] Cuny C, Strommenger B, Witte W, Stanek C. Clusters of infections in horses with MRSA ST1, ST254, and ST398 in a veterinary hospital. Microb Drug Resist. 2008;14(4):307–10.

[223] Moodley A, Stegger M, Bagcigil AF, Baptiste KE, Loeffler A, Lloyd DH, et al. spa typing of methicillin-resistant *Staphylococcus aureus* isolated from domestic animals and veterinary staff in the UK and Ireland. J Antimicrob Chemother. 2006;58(6):1118–23.

[224] Kock R, Ballhausen B, Bischoff M, Cuny C, Eckmanns T, Fetsch A, et al. The impact of zoonotic MRSA colonization and infection in Germany. Berl Munch Tierarztl Wochenschr. 2014;127(9–10):384–98.

[225] Tokateloff N, Manning ST, Weese JS, Campbell J, Rothenburger J, Stephen C, et al. Prevalence of methicillin-resistant *Staphylococcus aureus* colonization in horses in Saskatchewan, Alberta, and British Columbia. Can Vet J. 2009;50(11):1177–80.

[226] O'Mahony R, Abbott Y, Leonard FC, Markey BK, Quinn PJ, Pollock PJ, et al. Methicillin-resistant *Staphylococcus aureus* (MRSA) isolated from animals and veterinary personnel in Ireland. Vet Microbiol. 2005;109(3–4):285–96.

[227] van Duijkeren E, Moleman M, Sloet van Oldruitenborgh-Oosterbaan MM, Multem J, Troelstra A, Fluit AC, et al. Methicillin-resistant *Staphylococcus aureus* in horses and horse personnel: an investigation of several outbreaks. Vet Microbiol. 2010;141(1–2):96–102.

[228] Axon JE, Carrick JB, Barton MD, Collins NM, Russell CM, Kiehne J, et al. Methicillin-resistant *Staphylococcus aureus* in a population of horses in Australia. Aust Vet J. 2011;89(6):221–5.

[229] Van den Eede A, Martens A, Lipinska U, Struelens M, Deplano A, Denis O, et al. High occurrence of methicillin-resistant *Staphylococcus aureus* ST398 in equine nasal samples. Vet Microbiol. 2009;133(1–2):138–44.

[230] Maddox TW, Clegg PD, Diggle PJ, Wedley AL, Dawson S, Pinchbeck GL, et al. Cross-sectional study of antimicrobial-resistant bacteria in horses. Part 1: prevalence of antimicrobial-resistant *Escherichia coli* and methicillin-resistant *Staphylococcus aureus*. Equine Vet J. 2012;44(3):289–96.

[231] Vengust M, Anderson ME, Rousseau J, Weese JS. Methicillin-resistant staphylococcal colonization in clinically normal dogs and horses in the community. Lett Appl Microbiol. 2006;43(6):602–6.

[232] Springer B, Orendi U, Much P, Hoger G, Ruppitsch W, Krziwanek K, et al. Methicillin-resistant *Staphylococcus aureus*: a new zoonotic agent? Wien Klin Wochenschr. 2009;121(3–4):86–90.

[233] Burton S, Reid-Smith R, McClure JT, Weese JS. *Staphylococcus aureus* colonization in healthy horses in Atlantic Canada. Can Vet J. 2008;49(8):797–9.

[234] Morcillo A, Castro B, Rodriguez-Alvarez C, Abreu R, Aguirre-Jaime A, Arias A. Descriptive analysis of antibiotic-resistant patterns of methicillin-resistant *Staphylococcus aureus* (MRSA) st398 isolated from healthy swine. Int J Environ Res Public Health. 2015;12(1):611–22.

[235] Mammina C, Cala C, Plano MR, Bonura C, Vella A, Monastero R, et al. Ventilator-associated pneumonia and MRSA ST398, Italy. Emerg Infect Dis. 2010;16(4):730–1.

[236] Morales G, Picazo JJ, Baos E, Candel FJ, Arribi A, Pelaez B, et al. Resistance to linezolid is mediated by the cfr gene in the first report of an outbreak of linezolid-resistant *Staphylococcus aureus*. Clin Infect Dis. 2010;50(6):821–5.

[237] Kehrenberg C, Cuny C, Strommenger B, Schwarz S, Witte W. Methicillin-resistant and -susceptible *Staphylococcus aureus* strains of clonal lineages ST398 and ST9 from swine carry the multidrug resistance gene cfr. Antimicrob Agents Chemother. 2009;53(2):779–81.

[238] Gomez-Sanz E, Kadlec K, Fessler AT, Zarazaga M, Torres C, Schwarz S. A novel fexA variant from a canine *Staphylococcus pseudintermedius* isolate that does not confer florfenicol resistance. Antimicrob Agents Chemother. 2013;57(11):5763–6.

[239] Toh SM, Xiong L, Arias CA, Villegas MV, Lolans K, Quinn J, et al. Acquisition of a natural resistance gene renders a clinical strain of methicillin-resistant *Staphylococcus aureus* resistant to the synthetic antibiotic linezolid. Mol Microbiol. 2007;64(6):1506–14.

[240] Mendes RE, Deshpande LM, Castanheira M, DiPersio J, Saubolle MA, Jones RN. First report of cfr-mediated resistance to linezolid in human staphylococcal clinical isolates recovered in the United States. Antimicrob Agents Chemother. 2008;52(6):2244–6.

[241] Kehrenberg C, Schwarz S, Jacobsen L, Hansen LH, Vester B. A new mechanism for chloramphenicol, florfenicol and clindamycin resistance: methylation of 23S ribosomal RNA at A2503. Mol Microbiol. 2005;57(4):1064–73.

[242] Long KS, Poehlsgaard J, Kehrenberg C, Schwarz S, Vester B. The Cfr rRNA methyltransferase confers resistance to phenicols, lincosamides, oxazolidinones, pleuromutilins, and streptogramin A antibiotics. Antimicrob Agents Chemother. 2006;50(7):2500–5.

[243] Slifierz MJ, Friendship RM, Weese JS. Methicillin-resistant *Staphylococcus aureus* in commercial swine herds is associated with disinfectant and zinc usage. Appl Environ Microbiol. 2015;81(8):2690–5.

[244] Schwarz S, Cardoso M, Wegener HC. Nucleotide sequence and phylogeny of the tet(L) tetracycline resistance determinant encoded by plasmid pSTE1 from *Staphylococcus hyicus*. Antimicrob Agents Chemother. 1992;36(3):580–8.

[245] Fessler AT, Kadlec K, Hassel M, Hauschild T, Eidam C, Ehricht R, et al. Characterization of methicillin-resistant *Staphylococcus aureus* isolates from food and food products of poultry origin in Germany. Appl Environ Microbiol. 2011;77(20):7151–7.

[246] Argudin MA, Tenhagen BA, Fetsch A, Sachsenroder J, Kasbohrer A, Schroeter A, et al. Virulence and resistance determinants of German *Staphylococcus aureus* ST398 isolates from nonhuman sources. Appl Environ Microbiol. 2011;77(9):3052–60.

[247] Kadlec K, Schwarz S. Identification of a plasmid-borne resistance gene cluster comprising the resistance genes erm(T), dfrK, and tet(L) in a porcine methicillin-resistant *Staphylococcus aureus* ST398 strain. Antimicrob Agents Chemother. 2010;54(2):915–8.

[248] Schwendener S, Perreten V. New transposon Tn6133 in methicillin-resistant *Staphylococcus aureus* ST398 contains vga(E), a novel streptogramin A, pleuromutilin, and lincosamide resistance gene. Antimicrob Agents Chemother. 2011;55(10):4900–4.

[249] Kadlec K, Schwarz S. Novel ABC transporter gene, vga(C), located on a multiresistance plasmid from a porcine methicillin-resistant *Staphylococcus aureus* ST398 strain. Antimicrob Agents Chemother. 2009;53(8):3589–91.

[250] Kadlec K, Pomba CF, Couto N, Schwarz S. Small plasmids carrying vga(A) or vga(C) genes mediate resistance to lincosamides, pleuromutilins and streptogramin A antibiotics in methicillin-resistant *Staphylococcus aureus* ST398 from swine. J Antimicrob Chemother. 2010;65(12):2692–3.

[251] Hauschild T, Fessler AT, Kadlec K, Billerbeck C, Schwarz S. Detection of the novel vga(E) gene in methicillin-resistant *Staphylococcus aureus* CC398 isolates from cattle and poultry. J Antimicrob Chemother. 2012;67(2):503–4.

[252] Fessler AT, Kadlec K, Schwarz S. Novel apramycin resistance gene apmA in bovine and porcine methicillin-resistant *Staphylococcus aureus* ST398 isolates. Antimicrob Agents Chemother. 2011;55(1):373–5.

[253] Kadlec K, Fessler AT, Couto N, Pomba CF, Schwarz S. Unusual small plasmids carrying the novel resistance genes dfrK or apmA isolated from methicillin-resistant or -susceptible staphylococci. J Antimicrob Chemother. 2012;67(10):2342–5.

[254] Seier-Petersen MA, Nielsen LN, Ingmer H, Aarestrup FM, Agerso Y. Biocide susceptibility of *Staphylococcus aureus* CC398 and CC30 isolates from pigs and identification of the biocide resistance genes, qacG and qacC. Microb Drug Resist. 2015;21(5):527–36.

[255] Gomez-Sanz E, Torres C, Lozano C, Fernandez-Perez R, Aspiroz C, Ruiz-Larrea F, et al. Detection, molecular characterization, and clonal diversity of methicillin-resistant *Staphylococcus aureus* CC398 and CC97 in Spanish slaughter pigs of different age groups. Foodborne Pathog Dis. 2010;7(10):1269–77.

[256] Verhegghe M, Pletinckx LJ, Crombe F, Vandersmissen T, Haesebrouck F, Butaye P, et al. Methicillin-resistant *Staphylococcus aureus* (MRSA) ST398 in pig farms and multispecies farms. Zoonoses Public Health. 2013;60(5):366–74.

[257] Silva NC, Guimaraes FF, Manzi MP, Junior AF, Gomez-Sanz E, Gomez P, et al. Methicillin-resistant *Staphylococcus aureus* of lineage ST398 as cause of mastitis in cows. Lett Appl Microbiol. 2014;59(6):665–9.

[258] Anukool U, O'Neill CE, Butr-Indr B, Hawkey PM, Gaze WH, Wellington EM. Meticillin-resistant *Staphylococcus aureus* in pigs from Thailand. Int J Antimicrob Agents. 2011;38(1):86–7.

[259] Li D, Wu C, Wang Y, Fan R, Schwarz S, Zhang S. Identification of multiresistance gene cfr in methicillin-resistant *Staphylococcus aureus* from pigs: plasmid location and integration into a staphylococcal cassette chromosome mec complex. Antimicrob Agents Chemother. 2015;59(6):3641–4.

[260] He W, Liu Y, Qi J, Chen H, Zhao C, Zhang F, et al. Food-animal related *Staphylococcus aureus* multidrug-resistant ST9 strains with toxin genes. Foodborne Pathog Dis. 2013;10(9):782–8.

[261] Chao G, Zhang X, Zhang X, Huang Y, Xu L, Zhou L, et al. Phenotypic and genotypic characterization of methicillin-resistant *Staphylococcus aureus* (MRSA) and methicillin-susceptible *Staphylococcus aureus* (MSSA) from different sources in China. Foodborne Pathog Dis. 2013;10(3):214–21.

[262] Yan X, Yu X, Tao X, Zhang J, Zhang B, Dong R, et al. *Staphylococcus aureus* ST398 from slaughter pigs in northeast China. Int J Med Microbiol. 2014;304(3–4):379–83.

[263] Ho J, O'Donoghue M, Guardabassi L, Moodley A, Boost M. Characterization of methicillin-resistant *Staphylococcus aureus* isolates from pig carcasses in Hong Kong. Zoonoses Public Health. 2012;59(6):416–23.

[264] Brennan GI, Abbott Y, Burns A, Leonard F, McManus BA, O'Connell B, et al. The emergence and spread of multiple livestock-associated clonal complex 398 methicillin-resistant and methicillin-susceptible *Staphylococcus aureus* strains among animals and humans in the Republic of Ireland, 2010–2014. PLoS One. 2016;11(2):e0149396.

[265] Franco A, Hasman H, Iurescia M, Lorenzetti R, Stegger M, Pantosti A, et al. Molecular characterization of spa type t127, sequence type 1 methicillin-resistant *Staphylococcus aureus* from pigs. J Antimicrob Chemother. 2011;66(6):1231–5.

[266] Tavakol M, Riekerink RG, Sampimon OC, van Wamel WJ, van Belkum A, Lam TJ. Bovine-associated MRSA ST398 in the Netherlands. Acta Vet Scand. 2012;54:28.

[267] Porrero MC, Wassenaar TM, Gomez-Barrero S, Garcia M, Barcena C, Alvarez J, et al. Detection of methicillin-resistant *Staphylococcus aureus* in Iberian pigs. Lett Appl Microbiol. 2012;54(4):280–5.

[268] Lo YP, Wan MT, Chen MM, Su HY, Lauderdale TL, Chou CC. Molecular characterization and clonal genetic diversity of methicillin-resistant *Staphylococcus aureus* of pig origin in Taiwan. Comp Immunol Microbiol Infect Dis. 2012;35(6):513–21.

[269] Vestergaard M, Cavaco LM, Sirichote P, Unahalekhaka A, Dangsakul W, Svendsen CA, et al. SCCmec type IX element in methicillin resistant *Staphylococcus aureus* spa Type t337 (CC9) isolated from pigs and pork in Thailand. Front Microbiol. 2012;3:103.

[270] Van den Eede A, Hermans K, Van den Abeele A, Flore K, Dewulf J, Vanderhaeghen W, et al. The nasal vestibulum is the optimal sampling site for MRSA screening in hospitalised horses. Vet J. 2013;197(2):415–9.

[271] Rubin JE, Chirino-Trejo M. Antimicrobial susceptibility of canine and human *Staphylococcus aureus* collected in Saskatoon, Canada. Zoonoses Public Health. 2011;58(7):454–62.

[272] Anderson ME, Lefebvre SL, Rankin SC, Aceto H, Morley PS, Caron JP, et al. Retrospective multicentre study of methicillin-resistant *Staphylococcus aureus* infections in 115 horses. Equine Vet J. 2009;41(4):401–5.

[273] Zhang W, Hao Z, Wang Y, Cao X, Logue CM, Wang B, et al. Molecular characterization of methicillin-resistant *Staphylococcus aureus* strains from pet animals and veterinary staff in China. Vet J. 2011;190(2):e125–9.

[274] Vincze S, Stamm I, Kopp PA, Hermes J, Adlhoch C, Semmler T, et al. Alarming proportions of methicillin-resistant *Staphylococcus aureus* (MRSA) in wound samples from companion animals, Germany 2010–2012. PLoS One. 2014;9(1):e85656.

[275] Boost MV, O'Donoghue MM, Siu KH. Characterisation of methicillin-resistant *Staphylococcus aureus* isolates from dogs and their owners. Clin Microbiol Infect. 2007;13(7):731–3.

[276] Leonard FC, Abbott Y, Rossney A, Quinn PJ, O'Mahony R, Markey BK. Methicillin-resistant *Staphylococcus aureus* isolated from a veterinary surgeon and five dogs in one practice. Vet Rec. 2006;158(5):155–9.

[277] Schwaber MJ, Navon-Venezia S, Masarwa S, Tirosh-Levy S, Adler A, Chmelnitsky I, et al. Clonal transmission of a rare methicillin-resistant *Staphylococcus aureus* genotype between horses and staff at a veterinary teaching hospital. Vet Microbiol. 2013;162(2–4):907–11.

[278] Tirosh-Levy S, Steinman A, Carmeli Y, Klement E, Navon-Venezia S. Prevalence and risk factors for colonization with methicillin resistant *Staphylococcus aureus* and other Staphylococci species in hospitalized and farm horses in Israel. Prev Vet Med. 2015;122(1–2):135–44.

[279] Ishihara K, Saito M, Shimokubo N, Muramatsu Y, Maetani S, Tamura Y. Methicillin-resistant *Staphylococcus aureus* carriage among veterinary staff and dogs in private veterinary clinics in Hokkaido, Japan. Microbiol Immunol. 2014;58(3):149–54.

[280] Tajima K, Sinjyo A, Ito T, Noda Y, Goto H, Ito N. Methicillin-resistant *Staphylococcus aureus* keratitis in a dog. Vet Ophthalmol. 2013;16(3):240–3.

[281] Jang Y, Bae D, Cho JK, Bahk GJ, Lim SK, Lee YJ. Characterization of methicillin-resistant *Staphylococcus* spp. isolated from dogs in Korea. Jpn J Vet Res. 2014;62(4):163–70.

[282] Pak SI, Han HR, Shimizu A. Characterization of methicillin-resistant *Staphylococcus aureus* isolated from dogs in Korea. J Vet Med Sci. 1999;61(9):1013–8.

[283] Aklilu E, Zakaria Z, Hassan L, Hui Cheng C. Molecular relatedness of methicillin-resistant *S. aureus* isolates from staff, environment and pets at University Veterinary Hospital in Malaysia. PLoS One. 2012;7(8):e43329.

[284] van Duijkeren E, Box AT, Heck ME, Wannet WJ, Fluit AC. Methicillin-resistant staphylococci isolated from animals. Vet Microbiol. 2004;103(1–2):91–7.

[285] Mustapha M, Bukar-Kolo YM, Geidam YA, Gulani IA. Phenotypic and genotypic detection of methicillin-resistant *Staphylococcus aureus* in hunting dogs in Maiduguri metropolitan, Borno State, Nigeria. Vet World. 2016;9(5):501–6.

[286] Coelho C, Torres C, Radhouani H, Pinto L, Lozano C, Gomez-Sanz E, et al. Molecular detection and characterization of methicillin-resistant *Staphylococcus aureus* (MRSA) isolates from dogs in Portugal. Microb Drug Resist. 2011;17(2):333–7.

[287] Couto N, Belas A, Tilley P, Couto I, Gama LT, Kadlec K, et al. Biocide and antimicrobial susceptibility of methicillin-resistant staphylococcal isolates from horses. Vet Microbiol. 2013;166(1–2):299–303.

[288] Sieber S, Gerber V, Jandova V, Rossano A, Evison JM, Perreten V. Evolution of multidrug-resistant *Staphylococcus aureus* infections in horses and colonized personnel in an equine clinic between 2005 and 2010. Microb Drug Resist. 2011;17(3):471–8.

[289] Davis JA, Jackson CR, Fedorka-Cray PJ, Barrett JB, Brousse JH, Gustafson J, et al. Carriage of methicillin-resistant staphylococci by healthy companion animals in the US. Lett Appl Microbiol. 2014;59(1):1–8.

[290] van Balen J, Kelley C, Nava-Hoet RC, Bateman S, Hillier A, Dyce J, et al. Presence, distribution, and molecular epidemiology of methicillin-resistant *Staphylococcus aureus* in a small animal teaching hospital: a year-long active surveillance targeting dogs and their environment. Vector Borne Zoonotic Dis. 2013;13(5):299–311.

[291] Coughlan K, Olsen KE, Boxrud D, Bender JB. Methicillin-resistant *Staphylococcus aureus* in resident animals of a long-term care facility. Zoonoses Public Health. 2010;57(3):220–6.

[292] Morris DO, Rook KA, Shofer FS, Rankin SC. Screening of *Staphylococcus aureus, Staphylococcus intermedius*, and *Staphylococcus schleiferi* isolates obtained from small companion animals for antimicrobial resistance: a retrospective review of 749 isolates (2003–04). Vet Dermatol. 2006;17(5):332–7.

[293] Rankin S, Roberts S, O'Shea K, Maloney D, Lorenzo M, Benson CE. Panton valentine leukocidin (*PVL*) toxin positive MRSA strains isolated from companion animals. Vet Microbiol. 2005;108(1–2):145–8.

[294] Tomlin J, Pead MJ, Lloyd DH, Howell S, Hartmann F, Jackson HA, et al. Methicillin-resistant *Staphylococcus aureus* infections in 11 dogs. Vet Rec. 1999;144(3):60–4.

Section 3

Staphylococcus aureus Clinically

Chapter 4

Effects of Electronic (e)-Cigarette Vapor on Staphylococcal Virulence: Are E-Cigarettes Safer than Conventional Cigarettes?

Laura E. Crotty Alexander, Shymaa Enany and Elisa McEachern

Additional information is available at the end of the chapter

http://dx.doi.org/10.5772/66154

Abstract

In recent years, electronic (e)-cigarettes have dramatically increased in popularity as an alternative to conventional cigarettes. Little is known about the effects of e-cigarette vapor (EV) on bacteria that colonize the nasopharynx, including methicillin-resistant *Staphylococcus aureus* (MRSA). As most cases of pneumonia can be traced to bacteria in a patient's nasopharynx, increased virulence in potential pathogens could have direct consequences clinically for these patients. And because bacterial colonizers are spread between humans, increased virulence in one subject has implications for the community. There is accumulating evidence that exposure to cigarette smoke (CS) increases the pathogenicity of MRSA, as well as its dampening effects on the host immune system. EV exposure has also been demonstrated to increase MRSA virulence both *in vitro* and in a murine model of pneumonia. In this chapter, we will compare the virulence changes reported in MRSA exposed to CS vs. those exposed to EV, as well as proposed mechanisms and therapeutic targets.

Keywords: electronic cigarettes, e-cigarette vapor, cigarette smoke, staphylococcal virulence, MRSA, pneumonia

1. Introduction

Smoking is a leading cause of preventable morbidity and mortality worldwide, with an estimated 6 million premature deaths per year (10% of all deaths globally) attributable to cigarette smoke (CS) exposure [1]. Health ramifications of smoking include cardiovascular disease, malignancy, and chronic obstructive pulmonary disease [1]. In addition, smokers have increased rates of serious infections of the respiratory tract.

Tobacco smoke has many known carcinogenic and inflammatory effects on the airways and lungs [2]. Each puff of a cigarette exposes the epithelial and immune cells of the airway to more than 7000 chemicals including toxins such as acrolein, formaldehyde, and benzo*(a)*pyrene, as well as nicotine, the compound responsible for the addictive potential of both conventional cigarettes and e-cigarettes [1]. In chronic smokers, this chemical bath induces an inflammatory response in the airway that simultaneously damages the structural integrity of the airways and reduces the ability of host cells to clear pathogens from the airway [1–3]. Bacterial colonizers of the airways, including pathogens such as methicillin-resistant *Staphylococcus aureus* (MRSA), are exposed to the same toxic cigarette smoke milieu.

A newly fashioned nicotine delivery device, known commercially as electronic (e)-cigarette or vape pen was introduced to the international market for the first time in 2007 [4]. E-cigarettes have experienced a dramatic surge in popularity in the last decade. Their use among youth is a pressing public health issue, with prevalence of use surpassing that of tobacco cigarettes [5]. About 13.4% of USA youth are current users of e-cigarettes, a higher rate than that of traditional tobacco use. This corresponds to 3 million middle and high school students, up from 2.46 million in 2014 [6]. Moreover, e-cigarettes were found to increase smoking initiation by more than 200% with a prediction to achieve a corresponding 6% increase in smoking prevalence by 2060 [7]. One out of every four current e-cigarettes users in Spain have never smoked before [7, 8].

E-cigarettes consist of a cartridge containing nicotine in a propylene glycol (PG) and/or vegetable glycerin (VG) solvent that is heated and then vaporized via an atomizer; this vapor is subsequently inhaled, resulting in similar nicotine delivery to the bloodstream as conventional cigarettes. E-cigarettes have been promoted as a safer alternative to tobacco products, with fewer adverse effects and as being efficient in smoking cessation. In this respect, a recent study conducted in 28 European countries suggested a positive public health impact for e-cigarette for the tobacco smokers accompanied by minimal potential harm due to use by never smokers [9]. However, this study based its findings on a little-harm model of e-cigarettes, which is not supported by scientific data. E-cigarette components have been shown to contain many hazards for the respiratory system. In 2009, the US FDA found detectable levels of toxic cancer-causing chemicals in e-cigarette cartridges [10]. Many other components such as aldehydes, metals, volatile organic compounds, phenolic compounds, polycyclic aromatic hydrocarbons, and tobacco alkaloids were identified in e-cigarette cartridges, refill solutions, and aerosols [11]. All of these hazardous components in turn will come in contact with respiratory colonizing pathogens, including *S. aureus,* and may promote pro-virulent changes.

S. aureus are Gram-positive cocci with the potential to cause serious infections including bacteremia, endocarditis, and pneumonia, as well as skin and soft tissue infections ranging from impetigo to necrotizing fasciitis. More than one-third of the general population is asymptomatically colonized with *S. aureus* on either a transient or persistent basis, most frequently in the nose, throat, and perineum [12–14]. These colonizers are frequently the source of host staphylococcal infections, and patients with positive nasal swabs for *S. aureus* are known to be at high risk of nosocomial and surgical site infections [15]. Persistent carriers frequently have a higher bacterial load and a corresponding increased risk of subsequent staphylococcal infection relative to intermittent carriers [15].

Smokers are more likely than nonsmokers to be colonized with *S. aureus,* defined as >500 colony forming units (CFU) detected, with rates of colonization increasing with duration and frequency of smoking [14, 16, 17]. Smokers have an increased incidence of invasive staphylococcal infections than nonsmokers and have higher mortality as well [17–19]. Additionally, smokers are at greater risk of being colonized with antibiotic resistance *S. aureus* (MRSA) than nonsmokers. This finding is of note because MRSA infections have higher morbidity and mortality than methicillin-sensitive strains [20, 21]. Accumulating evidence suggests that this susceptibility to MRSA/staphylococcal infection may reflect not only the impairment of host defenses induced by chronic exposure to cigarette smoke, but also provirulent changes in commensal MRSA itself as a consequence of the same chronic cigarette smoke exposure.

To our knowledge, there are no epidemiologic studies of staphylococcal infection in e-cigarette users. However, recent work from our laboratory suggests that e-cigarette vapor (EV) may induce a similar stress response as conventional cigarette smoke in exposed *Staphylococcus,* resulting in a propathogenic phenotype.

2. Conventional cigarette smoke promotes increased staphylococcal virulence

MRSA often colonizes the nasopharynx, where it is exposed to everything that an individual inhales, including cigarette smoke (CS) or e-cigarette vapor (EV). Bacteria from the nasopharynx have the potential to travel to the lungs, causing pneumonia, or to the skin, where compromises in the epithelial barrier such as ulcers, abrasions, or surgical incisions provide ports of entry for invasive staphylococcal infection. The initial stage preceding infection is colonization of the human host. Transient colonization of the nasopharynx by *S. aureus* is more common in healthy individuals than persistent colonization (30 vs. 20%), which usually entails a larger bacterial burden and increased risk of infection [15]. Smokers are known to have higher rates of persistent colonization of the nasopharynx by MRSA relative to the general population [17]. The phenotypic changes in *S. aureus* induced by *in vitro* CS exposure favor persistent colonization via increasing adhesion to host cells and by inducing biofilm formation, reducing the ability of host immune effectors to limit colonization [22].

2.1. MRSA hydrophobicity changes induced by cigarette smoke

Our own studies demonstrate that CS extract (CSE) exposure results in a dose-dependent increase in MRSA surface hydrophobicity that is persistent and possibly even heritable [23, 24]. Increased hydrophobicity is associated with increased bacterial interactions with host epithelial cells [25]. Thus, we evaluated the adherence of MRSA to human keratinocytes (HaCaT cells) *in vitro* and found that pretreatment with CSE increased adherence from 28% (control) to 52% (CSE-MRSA). These findings are likely due to increased interaction with epithelial cell surfaces because of the increased surface hydrophobicity induced by CSE exposure. Increased capacity to adhere to epithelial cells as a consequence of CS exposure may explain the increased MRSA carriage rates in smokers.

2.2. Staphylococcal surface charge changes induced by cigarette smoke

CSE exposure induced dramatic changes in surface charge, in a dose-dependent manner. With increasing concentrations of CSE, the surfaces of *S. aureus* and MRSA became less negative. The negative surface charge of bacteria is protective against a variety of host defense mechanisms, including antimicrobial peptide (AMP) binding. Changes in surface charge were persistent for over 24 h post-CSE exposure. These changes suggest that the CSE exposure may result in persistent, heritable changes. Additionally, daily intermittent exposure to CSE over the course of 3 or 4 days—reflecting the pattern of exposure experienced by the nasopharyngeal flora of a smoker—resulted in an additive increase in surface charge alterations.

Cigarette smoke contains thousands of compounds, making the identification of specific etiologic agent inducing change in surface charge difficult. However, we hypothesized that reactive species such as free radicals and carbon monoxide likely played a role. To assess the contribution of reactive species, CSE was stored for 24 h at 4°C to allow time for volatile gases to evaporate and for some degradation of reactive oxygen species. The "aged" CSE induced a less-anionic surface charge in exposed MRSA; however, it had significantly less potency relative to fresh CSE. Thus, while reactive species account for some measure of the observed surface changes in MRSA, some of the more stable elements of CS contribute as well. One such component, present in both fresh and aged CSE—as well as e-cigarette vapor—is nicotine. We found that nicotine at 3 and 6 mg/mL induced a dose-dependent shift toward a less negative surface charge, suggesting that it likely plays a role in the altered surface charge of CSE-exposed MRSA. Finally, we sedimented CSE and exposed MRSA solely to these particulates and found no effect on surface charge.

The cell surface charge changes are mediated in part by expression of the *mprF* gene, which encodes a membrane protein involved in the shift to a less negative surface charge. Exposure to CSE induced a 1.8-fold increase in expression of *mprF* RNA, and CSE did not induce significant surface charge change in an *mprF* knock-out strain of *S. aureus* (SA113 Δ*mprF*), relative to wild-type bacteria, further supporting an important role of this pathway in this aspect of the staphylococcal response to CS exposure.

2.3. Cigarette smoke exposure increases *S. aureus* resistance to antimicrobial defenses

Aspiration of nasopharyngeal colonizers is a common occurrence and provides an opportunity for potential pathogens to cause airway infection. Numerous immunologic defenses are in place to protect against this outcome, including mechanical clearance of bacteria by the mucociliary elevator and destruction of remaining bacteria by immune cells within the airways. Alveolar macrophages, which destroy bacteria by mechanisms including phagocytosis, antimicrobial peptide (AMP) production, and respiratory burst, are an important component of host defense against pulmonary infection by bacteria.

http://dx.doi.org/10.5772/66154

CSE exposure may induce MRSA resistance to the bacteriocidal activities of these cells by multiple mechanisms. *In vitro*, CSE-exposed MRSA was resistant to killing by a murine alveolar macrophage cell line (MH-S), with a fourfold increase in survival relative to control ($p < 0.0001$). This outcome was not due to a cytotoxic effect on the macrophages, as macrophage cell death rate was comparable regardless of whether they were infected with CSE-treated MRSA or control bacteria. Phagocytosis of fluorescently tagged MRSA-GFP was unaffected by CSE exposure—suggesting that the difference in survival reflected that the CSE-exposed MRSA were better able to endure intracellular killing mechanisms including exposure to antimicrobial peptides and other toxic compounds and reactive oxygen species found in the phagolysosome. Indeed, CSE exposure increased resistance to killing by the human AMP LL-37, with twofold increases in both MIC and MBC relative to control bacteria. Decreased susceptibility to AMPs, which are produced by neutrophils, macrophages, and epithelial cells, suggests a significant increase in pathogenicity.

Potential mechanisms of this resistance include changes in cell surface charge and decreased rate of cell division induced by CS. Rapidly dividing bacteria are more sensitive to many antimicrobials, including conventional antibiotics and AMPs. Kristian et al. showed that growth suppression using bacteriostatic antibiotics allowed *S. aureus* decreased susceptibility to killing by AMPs [26]. Growth curves at various concentrations of CSE-demonstrated inhibition of bacterial growth in a CSE dose-dependent fashion. The growth defect induced by CSE resolved with removal of CSE; however, it is possible that a clonal population continues to divide slowly and that this population is better equipped to persist in the face of subsequent stressors.

Interestingly, the changes induced by CSE exposure also seemed to induce resistance to membrane solublization by detergent. CSE-MRSA had similar death rates to control bacteria during the first few minutes of incubation with detergent (Triton-X), but demonstrated improved survival subsequently, suggesting further protective cell membrane changes induced by CSE.

Findings *in vivo* in a murine model of pneumonia supported *in vitro* results, with CSE exposure resulting in increased staphylococcal persistence in the lungs and an increase in overwhelming infection. Mice infected with CSE-exposed MRSA had increased bacterial burdens in the lungs at 8 and 24 h relative to those infected with control bacteria, demonstrating that CSE-MRSA are better able to persist in the face of pulmonary immune defenses. Additionally, 40% of mice infected with a higher inoculum of CSE-exposed MRSA died within 48 h, vs. only 10% of those infected with unexposed MRSA.

2.4. Potential consequences of virulence changes induced by cigarette smoke

While pulmonary infection with MRSA is devastating, much of the morbidity and mortality associated with MRSA infections is due to soft tissue infections. Because bacteria colonizing the nasopharynx do not necessarily remain there, propathogenic changes induced by CS in the nasopharynx are of concern for staphylococcal infections beyond the pulmonary system. Coughing, sneezing, etc. lead to significant opportunity to transfer MRSA from nasal passages and upper airway to the skin, where an increased capacity to adhere to and invade epithelial cells—as well as resistance to AMPs—reflect serious potential consequences of the pro-pathogenic effects of cigarette smoke for skin and soft tissue infections. Finally, MRSA are transmitted between humans in the community as well. Thus, more virulent strains generated by CS exposure may put the community at higher risk of invasive staphylococcal diseases.

3. E-cigarette vapor increases staphylococcal virulence and impairs innate immune function

As the popularity of e-cigarettes skyrockets, it becomes increasingly important to establish whether e-cigarettes are indeed a safer alternative to conventional cigarettes. Simply, it is a battery operated device designed to deliver nicotine with flavors and other chemicals to users in vapor form instead of smoke. E-cigarette vapor (EV) contains a far more limited array of chemicals than cigarette smoke, resulting from vaporization of e-liquid containing nicotine, propylene glycol (PG), and vegetable glycerin (VG) solvents, with different flavors added in some cases. While PG is not harmful when ingested through the GI tract and is generally recognized as safe for use as a food additive [27], high wattage heating of these solvents results in the production of formaldehyde, aldehyde, and acrolein, resulting in significantly increased exposure of the airways to formaldehyde relative to a pack-day smoker [28]. The absorption of nicotine in blood showed no difference between the latest generation e-cigarettes and conventional cigarettes. Nicotine itself is toxic at high doses, and as described in the preceding section, induces a shift in MRSA toward a less negative surface charge which confers resistance to AMPs. Thus, e-cigarette "vaping" is exposing the colonizing bacteria to at least two components (formaldehyde and nicotine) with the potential to induce a significant stress response.

3.1. MRSA growth suppression by e-cigarette vapor

We tested the growth kinetics of MRSA cultures during exposure to different components of EV extract (EVE) including nicotine, PG, and VG. We also tested a number of different brands of e-cigarettes. EVE suppressed MRSA growth, with cultures failing to achieve logarithmic phase. While nicotine mildly inhibited MRSA growth in a dose-dependent fashion, the vaporized vehicles likely account for most of the suppressive effects observed. Vaporized PG and/or VG nearly abrogated MRSA growth, to the same degree as EVE [29]. Interestingly, MRSA growth suppression was observed in four of five brands, suggesting that the etiologic agent is a common component across brands. Thus, as with cigarette smoke, EV imposes significant stress on Staphylococcal cells. Bacterial survivors of the noxious exposure are likely to be hardier, and thus more difficult to kill.

3.2. Increased hydrophobicity, adherence, and invasion of keratinocytes by e-cigarette vapor exposed MRSA

As with conventional CS, EV promotes a shift in phenotype that supports persistent MRSA colonization of the epithelium, as well as increasing the risk of invasive infection. Hydrophobicity was markedly increased following EV exposure (by 31%), with a corresponding doubling of

MRSA adherence to the human keratinocyte HaCaT cell line. EVE-exposed MRSA demonstrated increased invasion of and intracellular persistence within HaCaT cells. These changes do not appear to be induced by nicotine, as increasing the nicotine content by fivefold had no effect on adherence and invasion.

3.3. E-cigarette vapor increases MRSA resistance to human antimicrobial peptide LL-37

Increased survival of internalized MRSA suggests that e-cigarette exposure may induce resistance to killing by epithelial cells. This effect is at least in part a consequence of moderately increased resistance to killing by antimicrobial peptides (AMPs). One of the established mechanisms for bacterial virulence is alterations in surface charge. Most bacteria have predominantly anionic surfaces, which are targeted by the human innate immune system [30, 31]. One of the human AMPs, LL-37, acts through charge interactions between its cationic surface and the bacterial lipids which are negatively charged [30–32]. EVE exposure increased the MIC of LL-37 from 7 uM in unexposed MRSA to 10 uM in EVE-MRSA ($P = 0.014$). As with conventional CS, EVE exposure induces a shift in MRSA toward a less negative surface charge. In turn, this reduces the propensity of the cationic LL-37 to bind the bacterial surface as it must for antimicrobial activity. This transition in surface charge was independent of the level of nicotine in EVE.

3.4. MRSA biofilm induction by e-cigarette vapor

EV has other pro-virulent effects on MRSA that may allow the bacteria to cause more severe infections—it promotes a moderate increase in biofilm formation. Antibiotics and our own immune cells have difficulty penetrating biofilm to kill embedded bacteria, thus allowing colonization (or infection) to perpetuate. The increased predilection for biofilm following EV exposure may reflect a response to nicotine in the vapor, as nicotine alone induced dose-dependent increases in biofilm formation. Interestingly, the increased biofilm formation occurring in response to conventional CS seemed to be due primarily to oxidative stress [29]. In our studies, MRSA incubated with only nicotine produced more biofilm than control. In the exposure to either EV or CS, a novel pathway in inducing MRSA biofilm was suggested [23, 29].

3.5. E-cigarette vapor increases MRSA virulence in a mouse model of pneumonia

In vivo studies in mice also suggested that EV promotes propathogenic features. The same murine pneumonia model used to evaluate CS effects on MRSA was employed to determine EV effects. Mortality was increased in mice infected with EV-MRSA (25 vs. 0% in mice infected with control MRSA, $p < 0.05$). In addition, bacterial burdens were 10-fold higher in mice infected with EV-MRSA relative to mice infected with MRSA controls. Exposure to EV enhanced both MRSA virulence and MRSA survival [29].

3.6. E-cigarette vapor induction of MRSA virulence gene expression

Similarly to the effect of CS exposure on MRSA virulence, MRSA exposed to EV was more virulent than controls in the biological model. Thus, the expression of several well-known virulence factors after EV exposure was evaluated. Panton-Valentine leukocidin (*pvl*), α-hemolysin (*hla*), coagulase (*coa*), α-phenol soluble modulin (*psm-α*), intracellular adhesion (*icaA*), staphylococcal protein A (*spa*), and quorum sensing (*agrA*) were quantified relative to 16s rRNA as a housekeeping gene. After EV exposure, the expression of *coa* and *pvl* increased by 1.68- and 1.56-fold, respectively [29]. Expression of *spa* did not change with EV exposure while *icaA, agrA, hla,* and *psm* decreased.

4. Cigarette smoke versus e-cigarette vapor

These data provide suggestive evidence that EV induces pro-virulent changes in MRSA comparable in type and magnitude to those promoted by exposure to conventional CS. Both EV and CS exposures resulted in shifts toward more hydrophobic and less anionic surface charges, with increased adherence and invasion of epithelial cells, and resistance to killing by the AMP LL-37. Additionally, the phenotypic changes induced by exposure to both inhalants result in significantly increased virulence in a mouse pneumonia model. Without running assays in parallel, however, definitive comparisons cannot be made (**Table 1**).

Both CS and EV exposure result in a similar profile based on these assays, likely reflecting a general stress response. However, the primary component of the inhalants inducing these

	Cigarette smoke		E-cigarette vapor	
Growth	↓↓	***	↓↓	****
Hydrophobicity	↑↑	****	↑	*
Adherence	↑	*	↑↑	**
Invasion	↑↑	**	↑↑	*
Surface charge changes	++++	****	+	****
Resistance to killing by:				
Macrophage	↑↑↑↑	****		
Antimicrobial peptide (AMP)	↑↑	*	↑	*
Cell lysis	↓	***		
Biofilm formation			↑↑	**
Virulence *in vivo*	↑↑	*	↑	*

*$p < 0.05$.
**$p < 0.01$.
***$p < 0.001$.
****$p < 0.0001$.

Table 1. Comparison of the effects of cigarette smoke and e-cigarette vapor on MRSA pathogenicity factors and virulence.

http://dx.doi.org/10.5772/66154

phenotypic changes may be different. This likely reflects that CS contains a multitude of toxic chemicals, with a more limited range found in EV. However, both nicotine and degradation products of the e-cigarette liquid solvents are clearly sufficient to induce concerning changes in MRSA *in vitro*. The effects of both CS and EV on MRSA in a physiologic setting may be synergistically negative due to the interplay of increased virulence with immunosuppressive effects of these substances on airway cells. Current studies are limited to animal models, and more research must be carried out to evaluate whether these changes reflect those observed following physiologic exposure in human users of both conventional and e-cigarettes. Additional avenues of research include evaluating the effects of exposure to these substances on staphylococcal soft tissue infections and bacteremia. In addition, epidemiologic studies on the incidence of staphylococcal infections in e-cigarette users are needed to assess the likelihood of our findings having a physiologic correlate.

Comparison of the effects of cigarette smoke and e-cigarette vapor on MRSA pathogenicity factors and virulence.

5. Possible downstream consequences of CS and EV on antibiotic resistance in MRSA

The effects of EV and CS on MRSA hydrophobicity, surface charge, virulence, and resistance to killing by the innate immune system host defense antimicrobial peptides have potentially harmful implications on antistaphylococcal therapy. Vancomycin and daptomycin are the only drugs approved by the US FDA for the treatment of MRSA bacteremia and endocarditis [33]. When MRSA causes endocarditis, it is due to the patient's colonizing strain. MRSA strains that are induced to relative resistance to cationic host defense peptides, such as platelet microbicidal proteins (PMPs), are associated with a proclivity to lead to persistent bacteremia, endocarditis, and metastatic infectious foci [34]. In other words, durable changes in MRSA physiology by EV and CS may have implications on the ability of this pathogen to establish endovascular infection in cases of an otherwise transient bacteremia that is thought to occur periodically in healthy hosts but otherwise rapidly quenched by the innate immune system. Furthermore, cross-resistance between host PMPs and cathelicidin and vancomycin and daptomycin have been well-documented [34–36]. Thus, EV and CS may have serious implications on the proclivity of MRSA to develop heteroresistance to these drugs, even before the agents are administered to the patient.

Compounding this, resistance concern is the fact that persistent lingering of MRSA *in vivo*, as a result of resistance to host defense peptide killing, further creates a selective pressure environment conducive to further cationic peptide resistance, and hence resistance to antimicrobial therapy [37]. These extrapolative hypotheses of the effects of EV and CS have yet to be proven, but if they are, have far greater public health implications than once thought.

Author details

Laura E. Crotty Alexander[1,2], Shymaa Enany[3*] and Elisa McEachern[4]

*Address all correspondence to: shymaa21@yahoo.com

1 Department of Medicine, University of California San Diego,San Diego, CA, USA

2 VA San Diego Healthcare System, San Diego, CA, USA

3 Department of Microbiology and Immunology, Faculty of Pharmacy, Suez Canal University, Ismailia, Egypt

4 Weill Cornell Medical College, New York, NY, USA

References

[1] National Center for Chronic Disease Prevention and Health Promotion, *Reports of the Surgeon General*, in *The Health Consequences of Smoking-50 Years of Progress: A Report of the Surgeon General*. 2014, Centers for Disease Control and Prevention (US): Atlanta (GA).

[2] Crotty Alexander, L.E., S. Shin, and J.H. Hwang, *Inflammatory diseases of the lung induced by conventional cigarette smoke: a review.* Chest, 2015. **148**(5): pp. 1307–22.

[3] Arcavi, L. and N.L. Benowitz, *Cigarette smoking and infection.* Arch Intern Med, 2004. **164**(20): pp. 2206–16.

[4] Rabe, K.F., et al., *Towards a total ban on links with the tobacco industry: new rules for the ERS.* Eur Respir J, 2012. **40**(4): pp. 809–10.

[5] Hammig, B., P. Daniel-Dobbs, and H. Blunt-Vinti, *Electronic cigarette initiation among minority youth in the United States.* Am J Drug Alcohol Abuse, 2016. DOI: 10.1080/00952990.2016.1203926.

[6] Arrazola, R.A., et al., *Tobacco use among middle and high school students—United States, 2011-2014.* MMWR Morb Mortal Wkly Rep, 2015. **64**(14): pp. 381–5.

[7] Cherng, S.T., et al., *Modeling the effects of e-cigarettes on smoking behavior: implications for future adult smoking prevalence.* Epidemiology, 2016. DOI: 10.1097/EDE.0000000000000497.

[8] Lidon-Moyano, C., et al., *Prevalence and user profile of electronic cigarettes in Spain (2014).* Gac Sanit, 2016. Doi: 10.1016/j.gaceta.

[9] Farsalinos, K.E., et al., *Electronic cigarette use in the European Union: analysis of a representative sample of 27 460 Europeans from 28 countries.* Addiction, 2016. Doi: 10.1111/add.13506.

[10] FDA, *FDA and Public Health Experts Warn About Electronic Cigarettes.* U.S. Food and Drug Administration, 2009. USA.

[11] Cheng, T., *Chemical evaluation of electronic cigarettes.* Tob Control, 2014. **23**(Suppl 2): pp. ii11–7.

[12] Armstrong-Esther, C.A., *Carriage patterns of Staphylococcus aureus in a healthy non-hospital population of adults and children.* Ann Hum Biol, 1976. **3**(3): pp. 221–7.

[13] Ridley, M., *Perineal carriage of Staphylococcus aureus.* Br Med J, 1959. **1**(5117): pp. 270–3.

[14] Sollid, J.U., et al., *Staphylococcus aureus: determinants of human carriage.* Infect Genet Evol, 2014. **21**: pp. 531–41.

[15] Wertheim, H.F., et al., *The role of nasal carriage in Staphylococcus aureus infections.* Lancet Infect Dis, 2005. **5**(12): pp. 751–62.

[16] Olsen, K., et al., *Staphylococcus aureus nasal carriage is associated with serum 25-hydroxyvitamin D levels, gender and smoking status. The Tromso Staph and Skin Study.* Eur J Clin Microbiol Infect Dis, 2012. **31**(4): pp. 465–73.

[17] Durmaz, R., et al., *Nasal carriage of methicillin-resistant Staphylococcus aureus among smokers and cigarette factory workers.* New Microbiol, 2001. **24**(2): pp. 143–7.

[18] Bagaitkar, J., D.R. Demuth, and D.A. Scott, *Tobacco use increases susceptibility to bacterial infection.* Tob Induc Dis, 2008. **4**: p. 12.

[19] Wooten, D.A. and L.G. Winston, *Risk factors for methicillin-resistant Staphylococcus aureus in patients with community-onset and hospital-onset pneumonia.* Respir Med, 2013. **107**(8): pp. 1266–70.

[20] Self, W.H., et al., *Staphylococcus aureus community-acquired pneumonia: prevalence, clinical characteristics, and outcomes.* Clin Infect Dis, 2016. **63**(3): pp. 300–9.

[21] Cosgrove, S.E., et al., *Comparison of mortality associated with methicillin-resistant and methicillin-susceptible Staphylococcus aureus bacteremia: a meta-analysis.* Clin Infect Dis, 2003. **36**(1): pp. 53–9.

[22] Kulkarni, R., et al., *Cigarette smoke increases Staphylococcus aureus biofilm formation via oxidative stress.* Infect Immun, 2012. **80**(11): pp. 3804–11.

[23] McEachern, E.K., et al., *Analysis of the effects of cigarette smoke on staphylococcal virulence phenotypes.* Infect Immun, 2015. **83**(6): pp. 2443–52.

[24] Crotty Alexander, L.E., et al., *Electronic cigarette vapor (ECV) exposure decreases Staphylococcus aureus susceptibility to macrophage and neutrophil killing.* Am J Respir Crit Care Med, 2014. **189**: p. A6624.

[25] Rosenberg, M. and S. Kjelleberg, Hydrophobic interactions: role in bacterial adhesion. In: *Advances in Microbial Ecology*, K.C. Marshall, Editor. 1986, Springer US: Boston, MA. pp. 353–393.

[26] Kristian, S.A., et al., *Impairment of innate immune killing mechanisms by bacteriostatic antibiotics.* FASEB J, 2007. **21**(4): pp. 1107–16.

[27] FDA. *Food Additive Status List.* 2014; Available from: http://www.fda.gov/Food/IngredientsPackagingLabeling/FoodAdditivesIngredients/ucm091048.htm.

[28] Jensen, R.P., et al., *Hidden formaldehyde in e-cigarette aerosols.* N Engl J Med, 2015. **372**(4): pp. 392–4.

[29] Hwang, J.H., et al., *Electronic cigarette inhalation alters innate immunity and airway cytokines while increasing the virulence of colonizing bacteria.* J Mol Med (Berl), 2016. **94**(6): pp. 667–79.

[30] Wang, G., *Structures of human host defense cathelicidin LL-37 and its smallest antimicrobial peptide KR-12 in lipid micelles.* J Biol Chem,, 2008. **283**(47): pp. 32637–43.

[31] Neville, F., et al., *Lipid headgroup discrimination by antimicrobial peptide LL-37: insight into mechanism of action.* Biophys J,, 2006. **90**(4): pp. 1275–87.

[32] Neville, F., et al., *Electrochemical screening of anti-microbial peptide LL-37 interaction with phospholipids.* Bioelectrochemistry, 2007. **70**(2): p. 205–13.

[33] Liu, C., et al., *Clinical practice guidelines by the infectious diseases society of america for the treatment of methicillin-resistant Staphylococcus aureus infections in adults and children: executive summary.* Clin Infect Dis, 2011. **52**(3): pp. 285–92.

[34] Sakoulas, G., et al., *Reduced susceptibility of Staphylococcus aureus to vancomycin and platelet microbicidal protein correlates with defective autolysis and loss of accessory gene regulator (agr) function.* Antimicrob Agents Chemother, 2005. **49**(7): pp. 2687–92.

[35] Mishra, N.N., et al., *In vitro cross-resistance to daptomycin and host defense cationic antimicrobial peptides in clinical methicillin-resistant Staphylococcus aureus isolates.* Antimicrob Agents Chemother, 2011. **55**(9): pp. 4012–8.

[36] Mishra, N.N., et al., *Reduced susceptibility to host-defense cationic peptides and daptomycin coemerge in methicillin-resistant Staphylococcus aureus from daptomycin-naive bacteremic patients.* J Infect Dis, 2012. **206**(8): pp. 1160–7.

[37] Kullar, R., J.A. McKinnell, and G. Sakoulas, *Avoiding the perfect storm: the biologic and clinical case for reevaluating the 7-day expectation for methicillin-resistant Staphylococcus aureus bacteremia before switching therapy.* Clin Infect Dis, 2014. **59**(10): pp. 1455–61.

Chapter 5

Staphylococcus aureus Bacteremia in Adults

Dima Youssef and Kate Molony

Additional information is available at the end of the chapter

http://dx.doi.org/10.5772/66225

Abstract

Staphylococcus aureus is an important cause of bacteremia, and *S. aureus* bacteremia constitutes a serious condition with high morbidity and mortality, secondary to multiple complications including infective endocarditis and embolization. The incidence of bacteremia with *S. aureus* is increasing with more frequent use of medications that lower immune system response, and with the utilization of more invasive medical procedures. In addition, the emergence of resistant *S. aureus* isolates is becoming more common and can negatively affect the outcome of an individual if not diagnosed and managed properly. Health care workers encounter *S. aureus* bloodstream infections on a routine basis, and in certain situations, it becomes a very challenging infection to control. Because of the impact this entity has on health care costs and the increased use of resources, it is necessary to highlight the causes, clinical presentation, associated complications, and treatment measures. In this chapter, we will cover each of these points, with somewhat more emphasis on methicillin-resistant *S. aureus* that is prevalent in both community and hospital settings and is more commonly associated with worsening prognosis and higher mortality.

Keywords: *Staphylococcus aureus*, Bacteremia, Sepsis, methicillin susceptible, methicillin resistant, community acquired, hospital acquired

1. *Staphylococcus aureus* infections: introduction

Staphylococcus aureus is a Gram-positive staphylococci that can exist commensally with humans as a colonizer but can also exist as a pathogen. It is a major pathogen in bacteremia whether community acquired or hospital acquired. It has proven its versatility by continuing to be an important infectious pathogen that has contributed to increasing morbidity and mortality of patients over the years. Despite the advances in antibiotic therapy targeting this pathogen, *S. aureus* remains a multipotent organism that causes infection using toxin production and nontoxin-mediated pathways. This organism causes a wide array of infections, from a simple skin infection to more dangerous situations such as bacteremia, endocarditis, pneumonia, bone and joint infections, and many others that may jeopardize the life of the patient. Bacteremia is one major cause of morbidity in both the inpatient and outpatient setting, and *S. aureus* is notorious for causing invasive infections that lead to bacteremia.

Patients with *S. aureus* bacteremia can be at risk for many complications that may increase morbidity, with mortality rates of 20–40% that have been described. The higher the level of resistance, the higher the mortality rates. This is why methicillin-resistant *S. aureus* (MRSA) is expected to have a higher morbidity/morality, longer hospital stays, and higher health care costs when compared with methicillin-sensitive *S. aureus* (MSSA) bacteremia [1]. Also, in cases of infection with MRSA, there is a higher rate of treatment failure that may include death within 30 days of receiving therapy, persistent positive blood cultures for more than 10 days after therapy, or recurrence of septicemia within 60 days after finishing therapy.

2. *S. aureus* colonization

S. aureus is a part of the normal human flora; up to 50% of healthy individuals may be persistently colonized with it. Colonization with *S. aureus* can be persistent in up to 20% of cases, intermittent in 60%, and always absent in up to 20% of people. In a study performed on the general US population that looked at colonization rates in the nares with *S. Aureus*, it was found that the prevalence of MRSA colonization was 0.8% between 2001 and 2002, and went up to 1.5% between 2003 and 2004. The anterior nares is felt to be the major site of *S. aureus* colonization, but some people can be colonized with *S. aureus* outside the nares in areas such as the throat, axilla, inguinal region, and perirectal area. Several conditions may increase the rate of colonization such as diabetes mellitus, HIV infection, underlying skin diseases, and end stage renal disease requiring hemodialysis. Colonization typically precedes *S. aureus* infection. These conditions can place the subject at a higher risk of invasive staphylococcal infections such as bacteremia, which is why much of infection control and prevention efforts target colonization with *S. aureus.*

Nasal carriage of *S. aureus* colonization has been associated with the development of infections. A substantial proportion of cases of *S. aureus* bacteremia appear to be of endogenous origin as they originate from colonies in the nasal mucosa. This is one reason why strategies to prevent systemic *S. aureus* infections by eliminating nasal carriage need to be supported.

3. Epidemiology

Since methicillin-resistant *S. aureus* constitutes a major burden on health care systems we will focus mainly on it. There are several terms for classifying MRSA infections, namely bacteremia. The first category is the health care–associated MRSA (also called nosocomial) that occurs more than 48 hours into hospitalization. The second category is community-onset health care–associated MRSA, which includes two factions: (1) patients in whom infection occurs less than 48 hours into hospitalization and (2) patients in the community who have had a prior hospitalization in the last 12 months (including for surgery or dialysis) or those who are residents of long-term care facilities. The third category is community-associated MRSA infections occurring outside of health care settings among individuals who do not have prior health care exposures. Several outbreaks of MRSA have occurred in the community without exposure to health care facilities. This reflects a great change in the epidemiology of MRSA-related infections. Once solely a hospital pathogen and only seen among individuals with prior health care exposures, now MRSA is seen in populations without health care exposures. Poor hygiene conditions, close contact, contaminated material, and damaged skin were found to be some of the risk factors for spread of MRSA infection in the community. In the United States, the most common MRSA community-acquired strain is the USA300 strain based on pulsed-field gel electrophoresis. This community-based clone mostly causes skin and soft tissue infections, but it may cause more invasive infections such as bacteremia in 5–10% of people. This clone is causing more nosocomial infections as well.

Besides being an important cause of community-acquired bacteremia such as in cases of intravenous drug use leading to endocarditis, or cases of intravenous home infusion therapy, *S. aureus* is a leading cause of nosocomial bacteremia. It ranks second after coagulase negative staphylococci as a cause of primary bacteremia. In the hospital setting, a higher prevalence of methicillin-resistant isolates is seen. Most of the time, bacteremia develops from *S. aureus* strains colonizing the host; however, this infection can be transmitted through contact with other colonized individuals or contaminated surfaces such as hands of health care workers or environmental spaces. Spread of staphylococci in aerosols of respiratory secretions from colonized patients has also been reported.

4. *S. aureus* virulence factors leading to bacteremia

In observing individual responses to MRSA infection, some hosts become severely ill while others have only mild symptoms. It is unclear why certain factors are directly linked to this discrepancy in response. There are several virulence factors of *S. aureus* that may be structural and secreted products that could cause the pathogenesis of the disease with *S. aureus*. Microbial surface components recognizing adhesive matrix molecules (MSCRAMMs) are surface proteins that mediate adherence of *S. aureus* to host tissues. These molecules bind molecules belonging to different surfaces such as fibronectin, collagen, and fibrinogen. The MSCRAMMs help establish invasive and serious infections like endovascular infections, bone and joint infections, and prosthetic-device infections. **Figure 1** represents a schema of the structural and secreted products that *S. aureus* uses in order to achieve a high virulence level, and serious infections like blood stream infection. **Table 1** listed a few selected virulence factors [2].

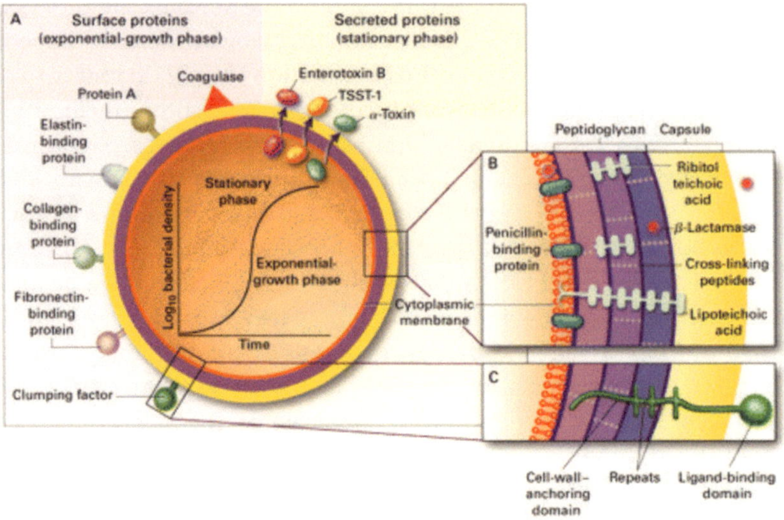

Figure 1. Pathogenic factors of *Staphylococcus aureus* with structural and secreted products both playing roles as virulence factors. (A) Surface and secreted proteins. (B and C) Cross sections of the cell envelope. TSST-1, toxic shock syndrome toxin-1. Source: With permission from the Massachusetts Medical Society. Copyright 1998 Massachusetts Medical Society.

Type of virulence factors	Selected factors[a]	Associated clinical syndromes
Involved in attachment	MSCRAMMs (e.g., clumping factors, fibronectin-binding proteins, collagen, and bone sialoprotein-binding proteins)	Endocarditis, osteomyelitis, septic arthritis, and prosthetic-device and catheter infections
Involved in persistence	Biofilm accumulation (e.g., polysaccharide intercellular adhesion), small-colony variants, and intracellular persistence	Relapsing infections, cystic fibrosis, and syndromes as described above for attachment
Involved in evading/ destroying host defenses	Leukocidins (e.g., PVL and γ-toxin), capsular polysaccharides (e.g., 5 and 8), protein A, CHIPS, Eap, and phenol-soluble modulins	Invasive skin infections and necrotizing pneumonia (CA-MRSA strains that cause these are often associated with PVL) abscesses (associated with capsular polysaccharides)
Involved in tissue invasion/penetration	Proteases, lipases, nucleases, hyaluronate lyase, phospholipase C, and metalloproteases (elastase)	Tissue destruction and metastatic infections
Involved in toxin-mediated disease and/or sepsis	Enterotoxins, toxic shock syndrome toxin-1, exfoliative toxins A and B, α-toxin, peptidoglycan, and lipoteichoic acid	Food poisoning, toxic shock syndrome, scalded skin syndrome, bullous impetigo, and sepsis syndrome
With poorly defined role in virulence	Coagulase, ACME, and bacteriocin	

Table 1. Selected *Staphylococcus aureus* virulence factors.

http://dx.doi.org/10.5772/66225

5. Pathogenicity

Several mechanisms lead to blood stream infection with *S. aureus*. After adhering to tissues or prosthetic materials, *S. aureus* is capable of growing in various ways. It can evade host defenses and the activity of antibiotics by forming biofilms on host and prosthetic surfaces. Additionally, *S. aureus* may escape the defense mechanisms by surviving inside several types of cells (such as endothelial cells) as in the situation of bacteremia and endocarditis. Another mechanism of survival is that *S. aureus* can form small-colony variants (SCVs) that can hide in host cells thus keeping them protected against defense mechanisms and leading to persistent and recurrent infection. The production of an antiphagocytic microcapsule is another method of defense escape used by *S. aureus* and can cause abscess formation. *S. aureus* can further halt host defenses by blunting neutrophil extravasation and chemotaxis to the infected area by producing chemotaxis inhibitory protein. Moreover, it produces leukocidins that destroy leucocytes by inflicting holes in the cell membrane.

Additional methods that help *S. aureus* in creating invasive blood stream infection exist and include the secretion of numerous enzymes that hydrolyzes tissues. This causes invasion, destruction and further spread of the pathogen to distant organs via the blood stream. Septic shock can thus result through the activation of the individual's immune system and coagulation pathways.

Pathogenesis of *S. aureus* is also affected by regulation of the expression of virulence factors. It appears that expression of these factors in a coordinated manner reduces the metabolic demands of the pathogen. Thus, MSCRAMM proteins that get secreted early in the infectious process help the establishment of the infection in tissue sites, while the later production of toxins facilitates the spread of the infection. The accessory gene regulator (*agr*) is a quorum-sensing system that plays a critical role in the regulation of staphylococcal virulence.

Besides virulence factors of *S. aureus*, it appears that patients were sicker when they developed an infection in the setting of negative colonization status. Noncarriers of the organism seem to have less protective immunity than those who are carriers. The formation of antibodies may also protect against the development of toxic shock syndrome.

Based on the above fact, *S. aureus* has many mechanisms to produce disease, namely bacteremia, while evading host defenses.

6. Bacteremia caused by *S. aureus*

Bacteremia is defined as the presence of bacteria in normally sterile blood. Typically more than one bottle in the set will be positive for growth; however, only one positive bottle is needed to diagnose bacteremia. Risk factors associated with *S. aureus* bacteremia include the presence of prosthetic devices, surgical site infections, or skin conditions such as chronic ulceration, injection drug use (IDU), and host factors that incur predisposition to recurrent infections. Prosthetic devices include any intravascular catheter such as hemodialysis catheter or central venous catheter. Patients on hemodialysis are at a higher risk for staphylococcal endocarditis and constitute a relatively new at-risk group. Other factors include defects of polymorphonuclear leukocytes and congenital syndromes that are associated with more risk of *S. aureus* infections, such as the cases of neutropenia, chronic granulomatous disease, as well as Job's, Chediak-Higashi, and Wiskott-Aldrich [3].

Clinical manifestations of *S. aureus* bacteremia typically involve systemic responses such as fever and hypotension. When bacteremia occurs secondarily to infection at a primary site, clinical symptoms associated with that organ system may also be present. Cellulitis, chronic ulceration, or trauma to skin and soft tissue may serve as portals of entry for the bacteria and the primary source of a *S. aureus* bacteremia. Tenderness or erythema surrounding a vascular catheter may also serve as a clinical manifestation of underlying bacteremia [4], though absence does not rule out the diagnosis. Patients with *S. aureus* pneumonia can develop bacteremia and have accompanying upper respiratory symptoms. *S. aureus* bacteriuria without the presence of a urinary catheter may be an indicator of *S. aureus* bacteremia [5]. *S. aureus* meningitis, though less common, may also occur in the setting of complication due to *S. aureus* bacteremia [6] and in addition to fever can demonstrate confusion and nuchal rigidity associated with acute bacterial meningitis.

Clinical approach to a patient with *S. aureus* bacteremia should include a detailed history, thorough physical exam, and if required, additional imaging with possible infectious disease consultation. History should involve questions as to the presence or absence of potential portals of entry such as wounds and also determine the presence of prosthetic devices including hardware (orthopedic or cardiac) and intravascular catheters. Questions related to localization of pain may help determine if metastatic spread has occurred such as in cases of vertebral osteomyelitis/diskitis or endocarditis. Physical exam should include an extensive evaluation of the skin and mucous membranes to look for sites of bacterial entry. Cardiac evaluation should assess for the presence of murmurs associated with infective endocarditis. Other stigmata of endocarditis should be sought through fundoscopic exam and exam of the digits for the appearance of emboli in skin. Baseline mental status should be noted and carefully monitored for signs of deterioration which may be concomitant with development of additional complications.

Complications of *S. aureus* bacteremia range from colonization after a treatment to infective endocarditis. Infective endocarditis is one of the most severe complications, with *S. aureus* now recognized as the most common cause in the industrialized world [7]. Pathogenesis is due to a combination of adhesion factors (as discussed earlier) on the surface of *S. aureus* and bacterial-induced platelet aggregation, which cause adhesion damage to heart valves [8]. Risk factors for IE in the setting of *S. aureus* bacteremia include prosthetic heart valve or predisposing cardiac abnormalities, IVDU, intravascular catheter infection, or persistent bacteremia [9]. Specific clinical manifestations associated with *S. aureus* infective endocarditis include sepsis syndrome involving fever, tachycardia, and hypotension, cardiac failure due to valve destruction, and sequelae from septic emboli. Within the heart, once *S. aureus* adheres to and colonizes the valve its intrinsic procoagulant activity triggers deposition of platelets and fibrin which leads to the formation of a vegetation. The structural abnormality is typically associated with regurgitation, and if untreated can progress to cardiac failure. Transthoracic echocardiography should be used as the initial diagnostic test in a patient with suspected endocarditis, as its specificity approaches 100% [10], however, specificity is lower being at most 75%. Transthoracic echocardiography is not 100% specific for infective endocarditis due to potential false positives, however, given a sensitivity greater than 90%, it is the better of the two for identification of valvular vegetations. Vascular phenomena occur when septic emboli dislodge from the vegetation and occlude arteries in the periphery as well as centrally affecting vital organs. Peripheral manifestations including skin lesions (Janeway spots, Osler's nodes) and retinal lesions (Roth's spots), while splenic vein thrombosis can lead to infarction of the spleen. Neurological complications include cerebral infarctions, intracerebral or sub-arachnoid hemorrhage, meningitis, cerebritis, and encephalomalacia.

7. Bacteremia treatment

Treatment of *S. aureus* bacteremia should first be approached by seeking out a potential focus of infection and determining whether or not it can be removed. Though no specific guidelines exist regarding duration of treatment, the general consensus advocates a 14-day treatment course for *S. aureus* bacteremia in cases where the source such as an intravascular catheter or prosthetic device can be removed, or an abscess can be drained [11]. In cases where removal of an intravascular catheter is not possible, antibiotic lock therapy may be used in an attempt to salvage the line, which includes filling the catheter lumen with high concentrations of antibiotics and leaving them in place for several hours to days [12]. Longer treatment courses extending for 4–6 weeks are required for deeper wound infections such as endocarditis and osteomyelitis. Methicillin-resistant *S. aureus* coverage should be included in empiric therapy with de-escalation to a beta-lactam agent if methicillin-susceptible *S. aureus* is later identified.

Once *S. aureus* susceptibility is determined, antibiotic therapy may be directed toward either MSSA or MRSA. Beta-lactams such as penicillins and cephalosporins, and if needed, glyco-peptides, are antibiotics classes used for the treatment of MSSA. Beta-lactams inhibit bacterial cell wall assembly by binding to membrane bound enzymes called penicillin-binding proteins that perform cross-linking. The beta-lactam ring binds to the penicillin-binding proteins and prevents the cross-linking component of cell wall assembly, causing cell death via autolysis of osmotic instability [13]. In cases where beta-lactams cannot be used to treat MSSA, such as with history of anaphylaxis to penicillin, the class of antibiotics known of as glycopeptides (which includes vancomycin) may be used. It should not be used as primary treatment for MSSA, however, if drug intolerance is not an issue.

Since MRSA bacteremia constitutes a great deal of infection in this day and constitutes a major cause of increasing morbidity and mortality, we decided to elaborate more about its treatment in different settings and to discuss the newer treatment options that are available.

8. Management of MRSA bacteremia and infective endocarditis in adults

MRSA was described in 1961, shortly after methicillin was introduced. Unlike penicillin resistance, which is achieved via the bacteria-produced enzyme penicillinase, methicillin resistance is mediated by a newly acquired penicillin-binding protein (called PBP2A) and encoded for by the *mecA* gene. The *MecA* gene is located on a mobile genetic element called staphylococcal chromosome cassette (SCCmec) [14]. If methicillin-resistant *S. aureus* bacteremia is identified, vancomycin and daptomycin are generally recommended for treatment based on current guidelines. Glycopeptides are a class of antibiotics that include vancomycin and work by binding to bacterial cell wall precursors and interfering with penicillin-binding protein enzymes, causing cessation of cell wall synthesis and later cell death. Daptomycin is a lipopeptide that is approved for the treatment of *S. aureus*-complicated skin or soft tissue infection, bacteremia and right-sided infective endocarditis [15]. Daptomycin diffuses through the peptidoglycan layer of Gram-positive organisms to the plasma membrane where it caused rapid depolarization resulting in the loss of membrane potential leading to loss of protein, DNA, and RNA synthesis and resulting in cell death [16].

In the case of uncomplicated bacteremia that is determined by the absence of endocarditis, artificial hardware, multiple sites of infection, and for which repeated blood cultures do not grow MRSA and patients are clinically well, vancomycin or daptomycin 6 mg/kg/dose IV once daily can be given for at least 2 weeks. However, in the case of complicated bacteremia, a duration of 4–6 weeks of therapy is recommended, depending on the extent of infection. Sometimes, higher dosages of daptomycin at 8–10 mg/kg/dose IV once daily may be needed.

When MRSA bacteremia becomes complicated with infective endocarditis, IV vancomycin or daptomycin 6–10 mg/kg/dose IV once daily for 6 weeks is recommended. It is not recommended to add gentamicin or rifampin to vancomycin for bacteremia or native valve infective endocarditis.

It is also important to identify the source and extent of the infection with removal and debridement or drainage of other sites of infection to decrease the bulk of the infection. Blood cultures need to be collected every 2–4 days after initial positive cultures until documentation of the clearance of bacteremia. And echocardiography is recommended for all adult patients with bacteremia to eliminate the possibility of associated endocarditis; transesophageal echocardiography (TEE) being preferred over transthoracic echocardiography (TTE). In the cases of large vegetations that exceed 10 mm in diameter, occurrence of more than one embolic event during the first 2 weeks of therapy, severe valvular insufficiency, valvular perforation or dehiscence, decompensated heart failure, perivalvular or myocardial abscess, new heart block, or persistent fevers or bacteremia, evaluation for replacement of the affected valve should be considered in consultation with cardiothoracic surgery.

In conditions that are characterized by MRSA bacteremia complicated with infective endocarditis of a prosthetic valve, administration of IV vancomycin plus rifampin 300 mg PO/IV every 8 h for at least 6 weeks plus gentamicin 1 mg/kg/dose IV every 8 h for 2 weeks is recommended, along with early evaluation for valve replacement surgery to decrease the risk of embolization.

9. Antimicrobial therapy that may be used for MRSA bacteremia

9.1. Clindamycin

Clindamycin is not specifically approved for treatment of MRSA infection, but it has been used for skin infections and invasive susceptible community-acquired MRSA infections in children. It is bacteriostatic and, as such, is not recommended for bacteremia, endovascular infections like infective endocarditis or septic thrombophlebitis. Clindamycin has excellent tissue penetration, particularly in bone and abscesses, but has poor penetration into the CSF. Community-acquired MRSA infections are more susceptible to Clindamycin than hospital-acquired isolates. It is important to have a D-zone test to look for inducible clindamycin resistance in erythromycin-resistant, clindamycin-susceptible isolates. Side effects include diarrhea and *Clostridium difficile*-associated disease. Clindamycin is pregnancy category B.

9.2. Daptomycin

This is a lipopeptide class antibiotic that destroys cell membrane function through calcium-dependent binding, leading in a bactericidal activity in a concentration-dependent manner. It is FDA approved for adults with *S. aureus* bacteremia, right-sided infective endocarditis, and complicated skin infections. It is not supposed to be given in nonhematogenous MRSA pneumonia because its activity is inhibited by pulmonary surfactant. The susceptibility breakpoint for daptomycin for *S. aureus* is ≤1 μg/mL. It appears that prior use of vancomycin and elevated vancomycin minimal inhibitory concentrations (MICs) has been associated with increases in daptomycin MICs and the emergence of nonsusceptible isolates. Monitoring creatinine phosphokinase (CPK) while on daptomycin is necessary to avoid rhabdomyolysis, which is seen with higher doses. Therapy with daptomycin may be complicated with daptomycin-induced eosinophilic pneumonia. Daptomycin is pregnancy category B.

9.3. Linezolid

Linezolid is a synthetic oxazolidinone and inhibits initiation of protein synthesis at the 50S ribosome. It is FDA-approved for treatment of skin infections and nosocomial pneumonia due to MRSA. It has a 100% oral bioavailability. Resistance to linezolid is rare, but has been reported. An outbreak with linezolid-resistant and methicillin-resistant *S. aureus* in an intensive care unit has been reported in Spain. Resistance to linezolid was mediated by the *cfr* gene, as all isolates ended up carrying this gene. It is not approved for the treatment of MRSA bacteremia, although it has been used for this condition on several occasions. Long-term use is not advisable as it may be complicated with hematologic toxicity, thrombocytopenia, anemia, neutropenia, peripheral and optic neuropathy, and lactic acidosis. Peripheral and optic neuropathy may not be reversible. Since it is a weak, nonselective, reversible inhibitor of monoamine oxidase, it may cause serotonin syndrome in patients taking concurrent selective serotonin-receptor inhibitors. It is considered pregnancy category C.

9.4. Tedizolid

Tedizolid is an oxazolidinone drug. It has the advantage of oral and parenteral formulations, similar to linezolid. It was approved for the treatment of acute bacterial skin and skin structure infections in 2014 and is administered once daily. Its use in bacteremia has not been recommended at this point.

9.5. Quinupristin-dalfopristin

Quinupristin-dalfopristin is constituted of two streptogramin antibiotics and inhibits protein synthesis. It is FDA approved for skin and soft tissue infections in adults and children >16 years of age. It has been used as salvage therapy for invasive MRSA infections in the setting of vancomycin treatment failure. It can have several side effects such as arthralgias, myalgias, and infusion-related reactions that may limit its use. Quinupristin-dalfopristin is considered pregnancy category B.

9.6. Rifampin

Rifampin is bactericidal against *S. aureus* and achieves high intracellular levels and good penetration in biofilms. It cannot, however, be used as monotherapy and is recommended to be used in combination with another antibiotic. It can be given at doses ranging from 600 mg daily in a single dose or in two divided doses to 900 mg daily in two or three divided doses. Rifampin is usually used in the setting of a *S. aureus* hardware infection.

9.7. Telavancin

Telavancin is an intravenous lipoglycopeptide. It inhibits cell wall synthesis by binding to peptidoglycan chain precursors and causing cell membrane depolarization. It has bactericidal activity against MRSA, vancomycin intermediate *S. aureus* (VISA), and vancomycin-resistant *S. aureus* (VRSA). It is FDA approved for complicated skin and soft tissue infections in adults and is pregnancy category C. Nephrotoxicity was more commonly reported among patients treated with telavancin than among those treated with vancomycin, however, unlike vancomycin, there is no need to monitor telavancin levels in the serum. It may be given in bacteremia, but would be an off label use.

9.8. Tetracyclines

Doxycycline is a tetracycline that is approved for the treatment of skin and soft tissue infections due to *S. aureus*. There is lack of data to support its use in more invasive infections like bacteremia. Tetracycline and doxycycline resistance in CA-MRSA is associated with *tetK* gene, but does not affect minocycline susceptibility. Minocycline is available in oral and parenteral formulations. A newer tetracycline named tigecycline is a glycylcycline and is a derivative of the tetracyclines. It is FDA approved in adults for skin and soft tissue infections and intraabdominal infections. It has a bacteriostatic activity against MRSA, thus it is not used in bacteremia; however, it was found that its use was associated with an increase in all-cause mortality. Tetracyclines are pregnancy category D and are not recommended for children <8 years of age due to the potential for tooth enamel discoloration and decreased bone growth.

9.9. Trimethoprim-sulfamethoxazole

TMP-SMX is not FDA-approved for the treatment of any staphylococcal infection, but since the majority of community-acquired MRSA strains are susceptible to it *in vitro*, it has become widely used for skin and soft tissue infections. It may also be used in bone and joint infections. For more invasive cases such as staphylococcal bacteremia and endocarditis, it can be used, though not as a first line drug. In addition, its use in the elderly must be done in conjunction

http://dx.doi.org/10.5772/66225

with close monitoring of creatinine and potassium levels. It is not recommended in pregnant women in the third trimester (pregnancy category C/D).

9.10. Ceftaroline

Ceftaroline is a fifth-generation cephalosporin. It is bactericidal against Gram-positive and Gram-negative pathogens and has activity against MRSA and VISA strains. It is recommended for skin and skin tissue infections and community-acquired pneumonia. Its use in cases of *S. aureus* bacteremia is still under investigation.

9.11. Dalbavancin

Dalbavancin is a semisynthetic lipoglycopeptide that inhibits cell wall synthesis. Its half-life is 147-258 hours, which allows use at once weekly dosing. It was approved in 2014 for treatment of acute bacterial skin and skin structure infections due to Gram-positive organisms, including MRSA. It is not yet approved for cases of *S. aureus* bacteremia.

9.12. Oritavancin

Oritavancin is a semisynthetic glycopeptide that also inhibits cell wall synthesis. Its half-life is 100 hours, allowing for single dose therapy. It was approved for treatment of acute bacterial skin and skin structure infections in 2014.

9.13. Vancomycin

Vancomycin has been the mainstay of parenteral therapy for MRSA infections; it has slow bactericidal activity. There is evidence of emerging resistant strains. Vancomycin kills staphylococci more slowly than β-lactams do *in vitro* and is inferior to β-lactams for MSSA bacteremia and infective endocarditis. Tissue penetration is highly variable and depends upon the degree of inflammation. Vancomycin's minimum inhibitory concentration breakpoints were changed in 2006 to improve the detection of intermediate susceptible strains (susceptible: MIC of 2 μg/mL or lower; intermediate: MIC of 4–8 μg/mL; and resistant: MIC 16 μg/mL or greater). The concept of MIC creep has arisen due to decrease in susceptibility to vancomycin among *S. aureus* isolates. *S. aureus* strains have been reported to "creep" up and approach the breakpoint of 2 with increasing frequency. This has been associated with worse clinical outcomes when vancomycin is used as therapy, when the MRSA isolate has a higher MIC to vancomycin. Vancomycin is considered pregnancy category C.

10. Management of persistent MRSA bacteremia and vancomycin treatment failures in adult patients

In cases of persistent positive blood cultures for *S. aureus*, it is necessary to look for deep-seated infections and hidden foci that continually send particles of infection into the blood stream. Removal of these infectious foci by either drainage or surgical debridement is recommended.

When vancomycin is used but the bacteremia persists, high-dose daptomycin (10 mg/kg/day), if the isolate is susceptible, in combination with another agent such as gentamicin 1 mg/kg IV every 8 h, rifampin 600 mg PO/IV daily, or 300–450 mg PO/IV twice daily, linezolid 600 mg PO/IV BID, TMP-SMX 5 mg/kg IV twice daily should be considered. But in case of reduced susceptibility to vancomycin and daptomycin, quinupristin-dalfopristin 7.5 mg/kg/dose IV every 8 h, TMP-SMX 5 mg/kg/dose IV twice daily, linezolid 600 mg PO/IV twice daily, or telavancin 10 mg/kg/dose IV once daily may be other options.

11. Recommendations for vancomycin dosing

In case of bacteremia, the dose of IV vancomycin is 15–20 mg/kg/day divided in two or three doses in order to conserve normal renal function. For seriously ill patients such as those with sepsis, meningitis, pneumonia, or infective endocarditis with suspected MRSA infection, a loading dose of 25–30 mg/kg (actual body weight) may be considered. Monitoring of vancomycin trough levels is necessary to guide the dosing of this antibiotic. Serum trough levels should be measured prior to the fourth or fifth dose. For serious infections such as bacteremia, infective endocarditis, meningitis, pneumonia, and necrotizing fasciitis due to MRSA, vancomycin trough concentrations of 15–20 μg/mL are recommended. Vancomycin trough monitoring is recommended for serious infections, patients who are morbidly obese have renal dysfunction or have fluctuating volumes of distribution. For isolates with a vancomycin MIC ≤ 2, the patient's clinical response should determine the continued use of vancomycin; however, if the patient has not had a clinical or microbiologic response to vancomycin despite adequate debridement and removal of other foci of infection, an alternative to vancomycin is recommended regardless of MIC. For the isolates with a vancomycin MIC >2 μg/mL (e.g., VISA or VRSA), an alternative to vancomycin should be used.

12. Prevention

Decolonization is important to achieve prevention of *S. aureus* bacteremia and other infections. The role of decolonization in controlling the spread of *S. aureus* is still unclear. It is also unclear what the optimal regimen is. Options include agents for nasal decolonization such as mupirocin and topical body decolonization with an agent such as chlorhexidine gluconate to target the extra nasal sites. Systemic oral antibiotics can be used for decolonization; however, there are issues that are very important to consider for decolonization, recolonization, and development of resistance. The current guidelines suggest that decolonization be considered in patients with recurrent skin infections or ongoing transmission occurring among household contacts despite optimizing wound care and hygiene measures. Hand hygiene consists of soap and water or an alcohol-based hand rub before and after contact with infected areas. Sharing personal items is discouraged.

As for hospitals, infection control and prevention strategies should include hand hygiene, active surveillance to identify *S. aureus* colonization, and environmental cleaning. Patient bathing with chlorhexidine gluconate in intensive care units leads to a reduction in *S. aureus* colonization and infection. It is felt that bathing with chlorhexidine gluconate is a measure for source control that may lead to less contamination of health care worker hands, thus less contamination of the environment and the spread of infection to other patients. One additional infection control strategy for years has been to create a vaccination against *S. aureus*. So far, attempts have been unsuccessful, but there is much research in this area.

13. Future perspective

S. aureus has had a steady increase in incidence over the last several decades. The higher frequency of artificial catheters, cardiac devices, joints being placed, of skin and surgical site wounds becoming infected, and intervenous drug use all serve as nidi for infection, particularly bacteremia. The cost and resource burden on health care systems is projected to continue to grow as the number of risk factors increase. There is also the problem to consider of how MRSA initially was only seen in health care settings but now makes up a large percentage of community-based infections.

What are some of the ways the medical community is working on not only treating but also preventing a much more widespread and resistant phenomenon? The approval of several newer antibiotics to combat serious MRSA infections shown in 2014, and there are a number of prospective antibiotics being studied with the potential to come to market [17]. A concerted effort among medical centers to make improvements at the level of the diagnostic stage (using transesophageal imaging more regularly) will be necessary in order to improve outcomes. In a different approach, the relationship among host immunologic factors in conjunction with environmental factors would be an additional avenue for exploration and possibly result in additional, nonantibiotic regimens. Continued use and awareness of infection prevention measures such as use of isolation inpatient and basic hand hygiene are both effective strategies in the greater attempt to not allow the bacteria to morph any and to prevent basic spread of the organism. Finally, there may be a time in the future when the ultimate means of infection control—a vaccination—would become available.

Author details

Dima Youssef* and Kate Molony

*Address all correspondence to: estecina@hotmail.com

Department of Internal Medicine, Division of Infectious Diseases, East Tennessee State University, Quillen College of Medicine, Johnson City, TN, United States

References

[1] Shurland S, Zhan M, Bradham DD, Roghmann MC. Comparison of mortality risk associated with bacteremia due to methicillin-resistant and methicillin-susceptible *Staphylococcus aureus*. Infect Control Hosp Epidemiol. 2007;28(3):273.

[2] Gordon RJ, Lowy FD. Pathogenesis of Methicillin-Resistant *Staphylococcus aureus* Infection. *Clinical infectious diseases : an official publication of the Infectious Diseases Society of America*. 2008;46(Suppl 5):S350-S359. doi:10.1086/533591.

[3] Moreillon P, Que YA, Glauser MP. *Staphylococcus aureus* (including staphylococcal toxic shock). In: Principles and Practice of Infectious Diseases, 6th ed., Mandell GL, Bennett JE, Dolin R (Eds), Churchill Livingstone, Philadelphia. 2005. p. 2321.

[4] Fowler VG Jr, Justice A, Moore C, Benjamin DK Jr, Woods CW, Campbell S, Reller LB, Corey GR, Day NP, Peacock SJ. Risk factors for hematogenous complications of intravascular catheter-associated *Staphylococcus aureus* bacteremia. Clin Infect Dis. 2005;40(5):695.

[5] Lee BK, Crossley K, Gerding DN. The association between *Staphylococcus aureus* bacteremia and bacteriuria. Am J Med. 1978;65(2):303.

[6] Tunkel AR, Hartman BJ, Kaplan SL, Kaufman BA, Roos KL, Scheld WM, Whitley RJ. Practice guidelines for the management of bacterial meningitis. Clin Infect Dis. 2004;39(9):1267.

[7] Fowler VG Jr, Miro JM, Hoen B, Cabell CH, Abrutyn E, Rubinstein E, Corey GR, Spelman D, Bradley SF, Barsic B, Pappas PA, Anstrom KJ, Wray D, Fortes CQ, Anguera I, Athan E, Jones P, van der Meer JT, Elliott TS, Levine DP, Bayer AS. *Staphylococcus aureus* endocarditis: a consequence of medical progress. JAMA. 2005;293:3012–3021. doi: 10.1001/jama.293.24.3012

[8] Fowler VG Jr, McIntyre LM, Yeaman MR, et al. In vitro resistance to thrombin-induced platelet microbicidal protein in isolates of *Staphylococcus aureus* from endocarditis patients correlates with an intravascular device source. J Infect Dis. 2000;182:1251–1254.

[9] Hill EE, Vanderschueren S, Verhaegen J, Herijgers P, Claus P, Herregods MC, Peetermans WE. Risk factors for infective endocarditis and outcome of patients with *Staphylococcus aureus* bacteremia. Mayo Clin Proc. 2007;82(10):1165.

[10] Habib G, Badano L, Tribouilloy C, Vilacosta I, Zamorano JL, Galderisi M, Voigt JU, Sicari R, Cosyns B, Fox K, Aakhus S, European Association of Echocardiography. Recommendations for the practice of echocardiography in infective endocarditis. Eur J Echocardiogr. 2010;11(2):202–219.

[11] Daneman N, Shore K, Pinto R, et al. Antibiotic treatment duration for bloodstream infections in critically ill patients: a national survey of Canadian infectious diseases and critical care specialists. Int J Antimicrob Agents. 2011;38:480–485.

http://dx.doi.org/10.5772/66225

[12] Mermel LA, Farr BM, Sherertz RJ, Raad II, O'Grady N, Harris JS, et al. Guidelines for the management of intravascular catheter-related infections. Clin Infect Dis. 2001;32:1249–1272.

[13] Novak R, Charpentier E, Braun JS, Tuomanen E. Signal transduction by a death signal peptide: uncovering the mechanism of bacterial killing by penicillin. Mol Cell. 2000;5(1):49.

[14] Chambers HF, Hartman BJ, Tomasz A. Increased amounts of a novel penicillin binding protein in a strain of methicillin-resistant *Staphylococcus aureus*. J Clin Invest. 1985;76:325–331.

[15] Fowler VG Jr, Boucher HW, Corey GR, et al. Daptomycin versus standard therapy for bacteremia and endocarditis caused by *Staphylococcus aureus*. N Engl J Med. 2006;355:653–665.

[16] Pogliano J, Pogliano N, Silverman JA. Daptomycin-mediated reorganization of membrane architecture causes mislocalization of essential cell division proteins. J Bacteriol. 2012;194(17):4494–4504. doi: 10.1128/JB.00011-12

[17] Rasmussen RV, Fowler VG Jr, Skov R, Brunn NE. Future challenges and treatment of *Staphylococcus aureus* bacteremia with emphasis on MRSA. Future Microbiol 2011. 6(1): 43–56.

Section 4

Detection of *Staphylococcus aureus*

Chapter 6

Determination of Staphylococcal Phenol-Soluble Modulins (PSMs) by a High-Resolution HPLC-QTOF System

Xiaoxiao Wu and Chi Zhang

Additional information is available at the end of the chapter

http://dx.doi.org/10.5772/66690

Abstract

Phenol-soluble modulins (PSMs) are multifunctional, amphipathic, α-helical peptides produced by virtually all staphylococcal strains. They have recently drawn much attention owing to the key contribution of some PSM peptides to staphylococcal virulence, in particular in highly virulent *Staphylococcus aureus*. High concentration of PSMs may cause cytolysis, damaging neutrophils, immune cells, red blood cells and white blood cells. Furthermore, all PSMs contribute to biofilm structuring and the dissemination of biofilm-associated infection. Here we describe a method for PSM analysis in *S. aureus* by high-performance liquid chromatography-quadrupole time-of-flight mass spectrometry system (HPLC-QTOF).

Keywords: phenol-soluble modulins, HPLC-QTOF, *Staphylococcus aureus*, determination

1. Introduction

Phenol-soluble modulins (PSMs) are a family of multifunctional, amphipathic and α-helical peptides produced by virtually all staphylococcal strains [1–3]. They have recently drawn much attention owing to the key contribution of some PSM peptides to staphylococcal virulence, in particular in highly virulent *Staphylococcus aureus* [4]. In addition, they have antibacterial activity, likely to compete with other environmental rival bacteria such as streptococci [5, 6], as well as biofilm structuring and dissemination functions [7, 8]. *Staphylococcus aureus* is an important and versatile opportunistic human pathogen that can cause a wide range of acute and chronic diseases, which range from superficial infections to invasive and life-threatening ones [9, 10].

PSMs were first identified in 1999 by the group of Seymour Klebanoff with the description of a "pro-inflammatory complex". They were isolated by hot phenol extraction from *S. epidermidis* culture filtrate [11]. The peptides were named PSMα, PSMβ and δ-toxin. Afterwards, the PSM compositon of *S.aureus* was also analyzed more systematically, which include the shorter α-type (four PSMα1-PSMα4 peptides and δ-toxin) and the longer β-type (the PSMβ1 and PSMβ2 peptides). The PSMs are encoded at three different locations in the genome. Four PSMα1-PSMα4 peptides are encoded in the PSMα operon. PSMβ1 and PSMβ2 are encoded in the PSMβ operon. δ-toxin is encoded within the coding sequence for RNAIII [12, 13].

S. aureus are considered to be second to salmonella as important foodborne pathogens, which have been frequently reported as agents leading to outbreaks of diseases caused by enterotoxins in ready-to-eat food and food products. The pathogenic mechanisms of *S. aureus* have been examined extensively through different stages of infection. Various staphylococcal components contribute to virulence with an improved understanding of specific functions [14–16].Proteomics studies have revealed that the production of virulence factors by different isolates of *S. aureus* is diverse and only a few of these seem to be invariantly produced [17]. PSMs are the most commonly identified staphylococcal virulence factors, especially in the community-associated (CA)-MRSA lineages. Recently, it was found that many *in vitro* PSM phenotypes were strongly inhibited by serum lipoproteins, and can exert their contribution to pathogenesis by intracellular killing and participate in neutrophil killing after phagocytosis [18]. In addition, PSMs at low concentration can lead to immune cell chemotaxis and inflammatory reaction. High concentration of PSMs may cause cytolysis, damaging neutrophils, immune cells, red blood cells and white blood cells [19, 20]. Thus, it is important to develop a rapid, specific and accurate method to detect PSMs in food products for the prevention of transmission.

At present, there are only a few methods for determination of PSMs, such as imaging mass spectrometry and liquid chromatography-ion trap or quadrupole mass spectrometry [11, 21]. The sample pre-treatment of the method using imaging mass spectrometry is quite cumbersome and time-consuming. The resolution of the method by liquid chromatography-ion trap or quadrupole mass spectrometry is too low to separate interfering substances with similar charge-to-mass ratios. Here, we describe a simple and effective method with higher sensitivity and selectivity for PSM analysis in *S. aureus* by high-performance liquid chromatography-quadrupole time-of-flight mass spectrometry system (HPLC-QTOF).

2. Materials

2.1. Reagents and chemicals

1. Tryptone soy broth (TSB).

2. *Staphylococcus aureus* strains.

3. 1-butanol.
4. Acetonitrile.
5. Formic acid.
6. Ultrapure water.
7. Eluent A: 0.1% formic acid in water, Eluent B: 0.1% formic acid in acetonitrile.

2.2. Instrument and equipment

1. Chromatograph: High-performance liquid chromatography-quadrupole time-of-flight mass spectrometry system (1260 HPLC system connected to a 6530 Accurate-Mass spectrometer Q-TOF/MS, Agilent, America).
2. Columns: Zorbax 300SB-C8 (2.1 μm, 4.6×150 mm, Agilent, America).
3. Mass Hunter software (Agilent, America).
4. Centrifuge (3-18K, Sigma, Germany).
5. Autoclave (DB-200, Systec, Germany).
6. Incubator shaker (ZWY-211B, ZhiCheng, China).
7. N-EVAP (Organomation, America).

3. Methods

3.1. Extraction of PSMs

Pre-cultures grown overnight from a liquid culture are used to inoculate bacterial cultures at 1:100 dilution.

3.1.1. Extraction of PSMs from TSB for qualitative analysis

1. The *Staphylococcus aureus* cells were removed by centrifugation for 10 min (5000 rpm, 4°C) after overnight growth (~16 h) in 100 mL of culture medium (TSB) at 37°C with shaking at 200 rpm.
2. Transfer supernatants to a 100-mL volumetric flask.
3. Add 1/3 (v/v) of 100% 1-butanol to the supernatants to make 25% (v/v) of 1-butanol.
4. Shake the solutions vigorously at 37°C for 2 h.
5. After brief centrifugation, collect upper (butanol) phases.
6. Dry the extract under nitrogen and redissolve the dried sample in ultrapure water.
7. Filter the solution through a 0.22-μm filter using a syringe before LC-MS analysis.

3.1.2. Extraction of PSMs from TSB and beverage for quantitative analysis

1. The *Staphylococcus aureus* cells were removed by centrifugation (5000 rpm, 10 min, 4°C) after overnight growth (~16 h) in 4 mL of culture medium (TSB or beverage) at 37°C with shaking at 200 rpm. The "beverage" means pure juice or juice beverage. Soda, water and coffee were not included.
2. Transfer 1.5 mL supernatants to 4-mL centrifuge tubes.
3. The supernatants were incubated at a 1:1 ratio with 1-butanol for 2 h (37°C, 200 rpm).
4. Transfer 1.0 mL supernatants to 2-mL centrifuge tubes.
5. Dry the extract under nitrogen and redissolve the dried sample in 0.5 mL ultrapure water.
6. Filter the solution through a 0.22-μm filter using a syringe before LC-MS analysis.

3.1.3. Extraction of PSMs from milk for quantitative analysis

1. The *Staphylococcus aureus* cells were precipitated by acetonitrile after overnight growth (~16 h) in 4 mL of culture medium (milk) at 37°C with shaking at 200 rpm.
2. Remove fat by frozen centrifugation technology (5000 rpm, 10 min, 0°C).
3. Extract water layer 1.5 mL to 4-mL centrifuge tubes.
4. The extract was incubated at a 1:1 ratio with 1-butanol for 2 h (37°C, 200 rpm).
5. Transfer 1.0 mL supernatants to 2-mL centrifuge tubes.
6. Dry the extract under nitrogen and redissolve the dried sample in 0.5 mL ultrapure water.
7. Filter the solution through a 0.22-μm filter using a syringe before LC-MS analysis.

3.1.4. Extraction of PSMs from pork for quantitative analysis

1. The *Staphylococcus aureus* cells were precipitated by acetonitrile after overnight growth (~16 h) in 4 mg of culture medium (pork) at 37°C with shaking at 200 rpm.
2. Add 1.5 mL 1-butanol and remove fat by frozen centrifugation technology (5000 rpm, 10 min, 0°C) after extraction of 2 h (37°C, 200 rpm).
3. Transfer 1.0 mL supernatants to 2-mL centrifuge tubes.
4. Dry the extract under nitrogen and redissolve the dried sample in 0.5 mL ultrapure water.
5. Filter the solution through a 0.22-μm filter using a syringe before LC-MS analysis.

3.2. Analysis of PSMs

3.2.1. Liquid chromatography methodology

1. Transfer the supernatants into glass sample vials. Cool down the sample tray to 10°C and put the sample vials in it.

2. An Agilent high-performance liquid chromatography-electrospray ionization-qudropole time-of-flight mass spectrometry (HPLC-ESI-QTOF/MS) system was used to analyze the PSMs qualitatively and quantitatively.

3. The chromatographic separation was performed with a Zorbax 300SB-C8 column (2.1 μm, 4.6×150 mm, Agilent) in series with a flow rate of 0.5 mL/min.

4. The gradient program can be described as follows: 10% eluent B for 2.5 min; 50% eluent B for 2.5 min; a linear gradient from 50% to 90 % eluent B for 15 min.

5. Injections (10 μL) were made using an autosampler, and the column temperature was set at 30°C.

3.2.2. Qualitative analysis of PSMs from culture

1. Transfer the supernatants into glass sample vials. Cool down the sample tray to 10°C and put the sample vials in it.

2. The ESI experiments were carried out by Dual AJS ESI ion source in a high-resolution instrument mode with a mass range of 3200 m/z and a positive scan mode with gas temperature at 325°C, drying gas flow rate at 5 L/min, sheat gas temperature at 350°C, and sheat gas flow rate at 7.5 L/min. The nebulizing gas was produced by a nitrogen generator.

3. The capillary voltage was kept at 4000 V and the fragmentor voltage was set to 175 V. Signals acquired in MS mode were used to make a preliminary identification, and data obtained in targeted MS/MS mode were searched from the online Mascot database, which could determine the PSMs accurately.

4. The peak area based on extracted ion chromatogram (EIC) can be used to quantitate the PSMs sensitively and rapidly.

4. Results and discussion

Examples of preset m/z ratios and retention times of *S. aureus* PSMs for qualitative and quantitative analyses are shown in **Table 1**. Retention times may vary according to the LC system and MS detector. (d), (t) and (q) stand for doubly, triply and quadruply charged ions, respectively.

The chromatograms of PSMs extracted from the culture are shown in **Figure 1**. The result indicates that the new method has a high selectivity. Good linearity of the PSMs was achieved in the range 0.5~100 μg/L ($R^2 > 0.99$). Compared with other analytical methods, the pre-treatment of the new method is simple and rapid, and the high resolution makes the method highly sensitive and selective. The method has been used successfully for the determination of PSMs extracted from different culture mediums, such as milk, beverage, vegetable and meat.

PSM	MS (m/z)	RT (min)	MS/MS (m/z)
PSMα1	1144.6842(d)	14.353	330.1491, 302.1539
PSMα2	1153.6964(d)	12.727	330.1475, 302.1538
PSMα3	1318.2228(d)	11.705	289.0861, 436.1542
PSMα4	1100.6982(d)	20.250	344.1674, 316.1711
δ-toxin	1002.8844(t)	15.010	474.1661, 587.2490
PSMβ1	1131.8610(q)	15.437	346.1059, 459.1892
PSMβ2	1121.8514(q)	11.353	318.1080, 702.3068

Table 1. m/z ratios and retention times of *S. aureus* PSMs.

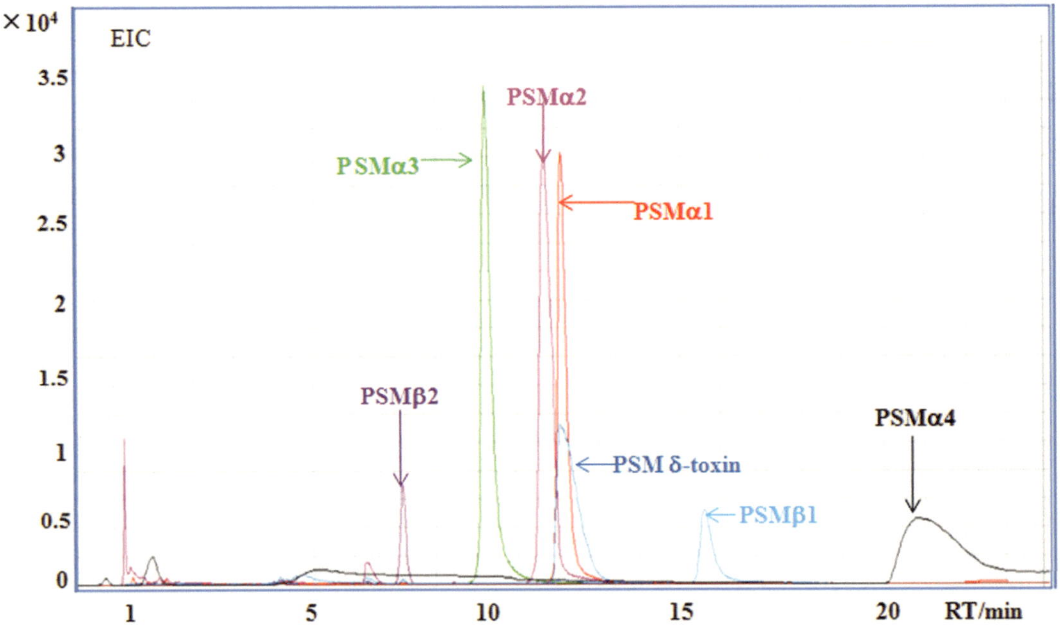

Figure 1. Chromatograms of PSMs extracted from the culture.

Acknowledgements

This work was supported by Science & Technology Project of China General Administration of Quality Supervision, Inspection and Quarantine (2015QK055), Science & Technology Project of Quality Supervision Bureau of Jiangsu Province (KJ155426), Six talent peaks project in Jiangsu Province (SWYY-031) and Jiangsu Natural Science Foundation (BK20141085).

Author details

Xiaoxiao Wu and Chi Zhang*

* Address all correspondence to: zhangchi@njzj.gov.cn

National Center of Supervision Inspection on Processed Food and Food Additives Quality, Nanjing Institute of Product Quality Inspection, Nanjing, China

References

[1] Gonzalez DJ, Vuong L, Gonzalez IS, Keller N, McGrosso D, Hwang JH, Hung J, Zinkernagel A, Dixon JE, Dorrestein PC. Phenol soluble modulin (PSM) variants of community-associated methicillin-resistant *Staphylococcus aureus* (MRSA) captured using mass spectrometry-based molecular networking. *Molecular & Cellular Proteomics*. 2014; **13**:1262–1272. doi:10.1074/mcp.M113.031336.

[2] Peschel A, Otto M. Phenol-soluble modulins and staphylococcal infection. *Nature Reviews Microbiology*. 2013; **11**:667–673. doi:10.1038/nrmicro3110.

[3] Cheung GYC, Joo HS, Chatterjee SS, Otoo M. Phenol-soluble modulins – critical determinants of staphylococcal virulence. *Fems Microbiology Reviews*. 2014; 38: 698–719. doi:10.1111/1574-6976.12057.

[4] Wang R, Braughton KR, Kretschmer D, Bach TH, Quech SY, Li M, Kennedy AD, Dorward DW, Klebanoff SJ, Peschel A, Deleo FR, Otto M. Identification of novel cytolytic peptides as key virulence determinants for community-associated MRSA. *Nature Medicine*. 2007; **13**:1510–1514. doi:10.1038/nm1656.

[5] Cogen AL, Yamasaki K, Sanchez KM, Dorschner RA, Lai YP, Macleod DT, Torpey JW, Otto M, Nizet V, Kim JE, Gallo RL. Selective antimicrobial action is provided by phenol-soluble modulins derived from *Staphylococcus epidermidis*, a normal resident of the skin. *Journal of Investigative Dermatology*. 2009; **130**:192–200. doi:10.1038/jid.2009.243.

[6] Joo HS, Cheung GY, Otto M. Antimicrobial activity of community-associated methicillin-resistant *Staphylococcus aureus* is caused by phenol-soluble modulin derivatives. *Journal of Biological Chemistry*. 2011; **286**:8933–8940. doi:10.1074/jbc.M111.221382. Epub 2011 Jan 28.

[7] Wang R, Khan BA, Cheung GY, Bach TH, Jameson-Lee M, Kong KF, Queck SY, Otto M. *Staphylococcus epidermidis* surfactant peptides promote biofilm maturation and dissemination of biofilm-associated infection in mice. *Journal of Clinical Investigation*. 2011; **121**:238–248. doi:10.1172/JCI42520.

[8] Periasamy S, Joo HS, Duong AC, Bach TL, Tan VY, Chatterjee SS, Cheung GYC, and Otto M. How *Staphylococcus aureus* biofilms develop their characteristic structure. *Proceedings of the National Academy of Sciences*. 2012; **109**:1281–1286. doi:10.1073/pnas.1115006109.

[9] Lowy FD. *Staphylococcus aureus* infections. *New England Journal of Medicine*. 1998; **339**:520–532. doi:10.1056/NEJM199808203390806.

[10] Peacock SJ, Silva ID, Lowy FD. What determines nasal carriage of *Staphylococcus aureus*? *Trends in Microbiology*. 2001; **9**:605–610. doi:10.1016/S0966-842X (01)02254-5.

[11] Mehlin C, Headley CM, Klebanoff SJ. An inflammatory polypeptide complex from *Staphylococcus epidermidis*: isolation and characterization. *Journal of Experimental Medicine*. 1999; **189**:907–918. doi:10.1084/jem.189.6.907.

[12] Queck SY, Khan BA, Wang R, Bach TH, Kretschmer D, Chen L, Peschel A, Deleo FR, Otto M. Mobile genetic element-encoded cytolysin connects virulence to methicillin resistance in MRSA. *PLoS Pathogens*. 2009; **5**:e1000533. doi:10.1371/journal.ppat.1000533.

[13] Chatterjee SS, Chen L, Joo HS, Cheung GY, Kreiswirth BN, Otto M. Distribution and regulation of the mobile genetic element-encoded phenol-soluble modulin PSM-mec in methicillin-resistant *Staphylococcus aureus*. *PLoS One*. 2011; **6**:e28781. doi:10.1371/journal.pone.0028781.

[14] Cheng AG, Kim HK, Burts ML, Krausz T, Schneewind O, Missiakas DM. Genetic requirements for *Staphylococcus aureus* abscess formation and persistence in host tissues. *Journal Official Publication of the Federation of American Societies for Experimental Biology*. 2009; **23**:3393–3404. doi:10.1096/fj.09-135467.

[15] Dumont AL, Nygaard TK, Watkins RL, Smith A, Kozhaya L, Kreiswirth BN, Shopsin B, Unutmaz D, Voyich JM, TorresVJ. Characterization of a new cytotoxin that contributes to *Staphylococcus aureus* pathogenesis. *Molecular Microbiology*. 2011; **79**:814–825. doi:10.1111/j.1365-2958.2010.07490.x.

[16] Cheung GYC, Duong AC, Otto M. Direct and synergistic hemolysis caused by Staphylococcus, phenol-soluble modulins: implications for diagnosis and pathogenesis. *Microbes & Infection*. 2012; **14**:380–386. doi:10.1016/j.micinf.2011.11.013.

[17] Breshears LM, Gillman AN, Stach CS, Schlievert PM, Peterson ML. Local Epidermal growth factor receptor signaling mediates the systemic pathogenic effects of *Staphylococcus aureus* toxic shock syndrome. *PLoS One*. 2016; **11**:e0158969. doi:10.1371/journal.pone.0158969.

[18] Surewaard BG, Nijland R, Spaan AN, Kruijtzer JA, de Haas CJ, van Strijp JA. Inactivation of staphylococcal phenol soluble modulins by serum lipoprotein particles. *PLoS Pathogens*. 2012; **8**:e1002606. doi:10.1371/journal.ppat.1002606.

[19] Otto M. Phenol-soluble modulins. *International Journal of Medical Microbiology*. 2014; **304**:164–169. doi:10.1016/j.ijmm.2013.11.019.

[20] Kretschmer D, Gleske AK, Rautenberg M, Wan R, Koberle M, Bohn E, Schoneberg T, Rabiet MJ, Boulay F, Klebanoff SJ, van Kessel KA, van-Strijp JA, Otto M, Peschel A.

Human formyl peptide receptor 2 senses highly pathogenic *Staphylococcus aureus*. *Cell Host Microbe*. 2010; **7**:463–473. doi:10.1016/j.chom.2010.05.012.

[21] Gonzalez DJ, Okumura CY, Hollands A, Kersten R, Akong-Moore K, Pence MA, Malone C, Derieux J, Moore BS, Horswill AR, Dixon JE, Dorrestein PC, Nizet V. Novel phenol-soluble modulin derivatives in community-associated methicillin-resistant *Staphylococcus aureus* identified through imaging mass spectrometry. *Journal of Biological Chemistry*. 2012; **287**:13889–13898. doi:10.1074/jbc.M112.349860.

Chapter 7

PCR Assay for Detection of *Staphylococcus aureus* in Fresh Lettuce (*Lactuca sativa*)

Andrés F. Chávez-Almanza, Jaime López-Cervantes,
Ernesto U. Cantú-Soto,
Dalia I. Sánchez-Machado and
Olga N. Campas-Baypoli

Additional information is available at the end of the chapter

http://dx.doi.org/10.5772/65559

Abstract

The growth in food demand and production growth of vegetables have led to the development of intensive production systems with the aim of having regular access to enough high-quality food. The aim is to determine the incidence of *Staphylococcus aureus* in fresh lettuce by PCR in order to enhance the efficiency for detection and identification process. The Baird-Parker method was used for isolating pathogens from 54 lettuce samples. Genomic DNA extraction was performed according the Mericon DNA Bacteria Plus Kit. The detection by PCR was performed using the pair of primers: *coa* gene (5′-ATAGAGCTGATGGTACAGG-3′ and 5′-GCTTCCGATTGTTCGATGC-3′). The phylogenetic tree was constructed by comparing conserved sequences from the adjacent 16S gene, using the F2C 5′-AGAGTTTGATCATGGCTC-3′ and C 5′-ACGGGCGGTGTGTAC-3′ primers. To test the antimicrobial effect, we used the disk diffusion method (Kirby-Bauer) using Mueller-Hinton agar and five antibiotics with different concentrations. The incidence of *S. aureus* was 1.7%. All the isolates were situated in the ATCC 11632 clade in accordance with other reported sequences belonging to this pathogen in the NCBI database. All the isolates seemed to be resistant to penicillin (10U). The molecular techniques used in this study are suitable for the identification of *S. aureus* isolated from lettuce, increasing our capability of detecting this pathogen by improving the process and increasing the efficiency contributing to the safety of this vegetable.

Keywords: *S. aureus*, fresh produce, lettuce, PCR, *coa* gene

1. Introduction

1.1. Taxonomy

Staphylococcus is a member of the *Micrococcaceae* family and consists of Gram-positive cocci, catalase-positive, that usually presents oxidative and fermentative metabolism of glucose. With measurements around 1 μm in diameter, their cells show a characteristic cluster of irregular arrangement as a result of cell division and some produce carotenoid pigments in yellow or golden colors [1, 2].

They are coagulase variant, non-spore-forming, facultative anaerobes (except *Staphylococcus saccharolyticus* reported to exhibit a faster and abundant development under aerobic conditions). *Staphylococcus* is catalase variant commonly positive (*Staphylococcus aureus* subesp. *anaerobius* and *S. saccharolyticus* are catalases negative), with Gram-variable capsule usually negative, (but always present), immobile, and oxidase variant commonly negative (*Staphylococcus caseolyticus, Staphylococcus lentus, Staphylococcus sciuri* and *Staphylococcus vitulus* are positive to the modified oxidase reaction). A O/F glucose test determines them as F (fermentative) microorganisms, with optimum growth temperature from 30 to 37°C. There are 44 species sensitive to lysis by lysostaphin, but resistant to lysozyme (*S. arlettae, S. aureus* subesp. *anaerobius, S. aureus* subesp. *aureus, S. auricularis, S. capitis* subesp. *capitis, S. capitis* subesp. *ureolyticus, S. caprae, S. carnosus, S. caseolyticus, S. chromogenes, S. cohnii* subesp. *cohnii, S. cohnii* subesp. *urealyticumk S. delphini, S. epidermidis, S. equorum, S. felis, S. gallinarium, S. haemoliticus, S. hominis, S. hyicus, S. hyicus* subesp. *hyicus, S. intermedius, S. kloosii, S. lentus, S. lugdunensis, S. lutrae, S. muscae, S. pasteuri, S. piscifermentans, S. pulvereri, S. sacharolyticus, S. saprophyticus* subesp. *bovis, S. saprophyticus* subesp. *saprophyticus, S. schleiferi, S. schleiferi* subesp. *coagulans, S. schleiferi* subesp. *schleiferi, S. sciuri, S. sciuri* subesp. *carnaticus, S. sciuri* subesp. *rodentium, S. sciuri* subesp. *sciuri, S. simulans, S. vitulus, S. warneri, S. xylosusk*). *S. aureus* subesp. *aureus, S. epidermidis, S. haemolyticus, S. lugdunesis, S. saprophyticus* subesp. *saprophyticus* are species of *Staphylococcus* isolated most frequently associated with human infections [1]).

S. aureus is considered a pathogen with high potential to cause multiple infections in humans and animals; it was discovered by Dr. Alexander Ogston in 1880 [3]. Being the most common pathogen causing nosocomial infections [4, 5], as well as cases of food poisoning [6], it is considered the most virulent bacteria, responsible for a broad spectrum of diseases, ranging from skin and soft tissue infections to serious diseases that threaten life. *S. aureus* belongs to the normal human flora, and 25–50% of the healthy population is colonized with this bacterium, constituting a risk of dissemination, as it can be acquired through contact with other people or through environmental exposure [3].

Pathogenicity of *S. aureus* infections is related to various components of the bacterial surface generally formed by peptidoglycan and teichoic acids, in addition to protein A [7]. The pathogenesis caused by this microorganism occurs when the combination of virulence factors with decreased host defenses occurs [8], and these conditions favor the microorganism having the characteristics of virulence and damage [9]. In addition, the situation is aggravated because this pathogen has developed resistance to multiple antibiotics, making the development of novel treatments and medication for diseases caused by this bacterium more challenging [10].

1.2. *S. aureus* in food

The presence of pathogens in minimally processed vegetables and their ability to survive and grow has been well documented [11]. *S. aureus* (including methicillin-resistant *S. aureus*, MRSA) is one of the most common pathogens in food and is considered responsible for the most common types of poisoning in meats, salads, milk, and dairy products [12–14]. A wide variety of foods can be a vehicle for *S. aureus* poisoning; besides pollution exposure (usually human) and the use of raw materials and contaminated ingredients (typically but not exclusively of animal origin), ecological conditions must be met in food to favor growth of the microorganism, including temperatures above 20–25°C, even for a few hours.

Therefore, it is considered that *S. aureus* has a remarkable ability to proliferate in various foods; under conducive conditions, the growth rate can lead to enough concentration of enterotoxins to cause severe outbreaks of gastroenteritis. Meat, dairy products, certain vegetables, and cooked foods work as excellent growth medium to support their multiplication. It is important to mention that not all foods are favorable substrate for the development of the microorganism; in fact, some are inhibitory; development patterns of *S. aureus* in some foods are favorable (raw meat, cooked chicken, fresh cheese, raw milk, yolk), unfavorable (raw vegetables, ground beef, raw fruit, dried fruit, nuts), and inhibitory (chocolate, cocoa, mature dairy, processed juice, fermented food) [12].

There are *S. aureus* strains more pathogenic than others: one example is the methicillin-resistant *S. aureus* (MRSA). Methicillin is a semi-synthetic derivative of penicillin introduced in Europe in 1959; one year after its introduction, the first strain of MRSA was detected, and in 1963, the first nosocomial outbreak caused by this microorganism was reported. MRSA has been thoroughly studied at a genetic level, due to its recurrent appearance in later years. In some cases, it has been associated with food consumption, and knowledge about the spread and epidemiology has been used to develop strategies to prevent the distribution of MRSA. Simultaneously, the development of various molecular typing techniques has emerged, aimed at detecting the phenotypes or specific molecular characteristics of each strain in question [15].

1.3. Staphylococcal intoxication

One of the most important foodborne diseases transmitted around the world is Staphylococcal intoxications; of all outbreaks of food poisoning that occur, on average 20% are due to the consumption of food contaminated with enterotoxins produced by bacteria of the genus *Staphylococcus* and mainly for the *S. aureus* species. The intoxication is characterized by nausea, vomiting, abdominal cramps, malaise, headache, and occasionally diarrhea without the presence of fever. Symptoms can appear 30 min after consumption of the aliment, with the most common incubation period going from 2 to 4 h [2].

Commonly, *S. aureus* has been isolated from plants [16]. Microbial contamination in food can occur due to poor storage procedures, and an increase in temperature during shelf life has been identified as the cause of microorganism proliferation [11]. Infections by Staphylococcal enterotoxins have constituted the leading cause of foodborne disease in the United States, and these enterotoxins are the leading cause of outbreaks caused by contaminated food in the European Union. In June 2000, Japan reported a mass intoxication of more than 10,000 cases caused by this organism present in milk [6, 17]. The Center for Disease Control and Prevention (US-CDC) reports that consumption of contaminated food with *S. aureus* causes 185,060 cases of infection, 1753 hospitalizations, and two deaths annually just in this country [13].

In the Middle East, many types of vegetables are eaten raw in salads or used as garnish appetizers, and in traditional meals, they are perceived as healthy food; however, in other parts of the world, these raw vegetables have been major contributors of foodborne diseases in recent years [18, 19]. In the United States, green leafy vegetables have been identified as part of the 10 riskiest foods regulated by the Food and Drug Administration (FDA), representing almost 40% of foodborne outbreaks according to data obtained from the Center for Disease Control and Prevention (CDC) [20].

The consumption of green leafy vegetables provides numerous health benefits, and there is a direct relationship between consumption of these vegetables and the reduction of chronic diseases such as hypertension, diabetes, atherosclerosis, and cancer [21]. Currently, most fresh-cut products are washed in chlorinated water (50–200 mg L^{-1} of active chlorine) to reduce the levels of microorganisms. Sodium hypochlorite (NaCl) is the most widely used disinfectant in the fresh-cut industry [22]. The lack of thorough cooking in fresh cuisine can result in foodborne diseases if contaminated by pathogens. Despite these foods being ready to eat, it has been reported that their quality is not satisfactory in Vienna, Austria [23], Johannesburg, South Africa [24], Korea [25], and Catalonia, Spain [26]. Reports show that the main pathogens in ready-to-eat foods include *Listeria monocytogenes*, *S. aureus*, *Bacillus cereus*, *Salmonella* spp., and *Escherichia coli* O157:H7, the last two being involved in most outbreaks caused by fresh fruits and vegetables [27, 28] reported in low doses of 10 and 2–2000 cells, respectively [29, 30].

In Mexico, staphylococcal intoxications are responsible for 45% of the outbreaks caused by food poisoning. These data partially reflect the incidence of this disease in the country, considering that it only represents the outbreaks that have been reported or studied; however, it is useful to show that on a national level, staphylococcal poisoning is a major foodborne [2]. The main places where the outbreaks were reported to occur are at parties or social gatherings, schools or daycare centers, restaurants, and hospitals, in this order of importance [31].

In recent years, Mexico has become one of the most dynamic markets for the US horticultural importers, displacing Japan to third place [32]. From January to April 2014, a total of 51,109 tons of lettuce equivalent to 42.289 mdd were exported, the main exporters being: Guanajuato (56.8%), Nuevo Leon (19.9%), Baja California (17%), Sonora (1.9%), other states (4.4%) [33].

1.4. Identification of *S. aureus*

Various methods have been developed for the isolation and quantitative identification of *S. aureus*; the selection of the method depends on the type of food, as well as the history of incidence of the pathogen.

Processed foods may contain relatively small numbers of debilitated viable cells, whose presence must be demonstrated by appropriate means (BAM, Bacteriological Analytical Manual) [34].

Among the most important methods used globally are those recommended by the Association of Official Analytical Chemistry (AOAC) (975.55-1976, *S. aureus* in foods. Surface plating) and the Food and Drug Administration (BAM, Bacteriological Analytical Manual Chapter 12: *S. aureus*). In Mexico, the microbiological method to determine the account of *S. aureus* present in national or imported foods is established in the Mexican Official Standard NOM-210 appendix B [35], in accordance with the International Standard ISO 6888-1:1999. Microbiology of food and animal feeding stuffs—horizontal method for the enumeration of coagulase-positive staphylococci (*S. aureus* and other species)—Part 1: Technique using Baird-Parker agar medium.

A brief description of the method includes:

1.4.1. Baird-Parker method

1.4.1.1. Preparation of the samples

Take different portions of the food, transfer 25 g or mL into dilution bottles with 225 mL of phosphate buffer or peptone water to prepare a dilution 1:10, and homogenize for 1 or 2 min in blender or peristaltic homogenizer.

1.4.1.2. Analytical procedure

Transfer 0.1 mL of direct sample with a sterile pipette if liquid, or 0.1 mL of the initial suspension (dilution 10^{-1}) in the case of other products, onto plates with Baird-Parker agar with addition of egg yolk emulsion. Do this in duplicates and repeat this procedure for subsequent dilutions 10^{-2}, 10^{-3} if necessary.

Carefully distribute the inoculum on the agar surface as soon as possible, with a sterile glass rod bent at a right angle, using one for each plate and dilution. The plates must be kept with the top upward until the inoculum is fully absorbed by the agar.

Invert and incubate the plates from 44 to 48 h at 36°C and subsequently search for colonies with typical morphology: black in color, circular, bright, convex, flat from 1 to 2 mm in diameter, showing one opaque zone, wet and with a clear halo.

Select the plates having between 15 and 20 typical and atypical colonies for their confirmation. From each sample, select five typical colonies for confirmation or five atypical colonies to perform Gram staining. In the case of observing positives bacilli, the colony will be taken as negative for *S. aureus;* on the contrary if cocci are observed, the confirmation will continue.

When the plates contain <15 typical colonies, a note citing "estimated value" must be added to the report of results.

1.4.1.3. Confirmation procedure

Coagulase test: Coagulase is a protein produced by various microorganisms that enables the conversion of fibrinogen to fibrin. In the laboratory, it is used to distinguish between different types of *S. aureus* [36]. Dehydrated rabbit plasma is used for the test (this is rehydrated following the manufacturer's instruction).

For the procedure, select and inoculate each typical colony in a tube with 0.5 mL brain heart infusion broth (BHI) and tubes with Trypticase soy agar (TSA). Simultaneously use a positive control (*S. aureus*) and a negative control (*S. epidermidis*). Incubate at 35 + 1°C in a water bath from 20 to 24 h. Keep the bacterial culture in ATS at temperature no more than 27°C for subsequent tests. Add 0.1 mL of the previous bacterial culture to 0.3 mL of rabbit plasma with EDTA (unless the manufacturer indicates other quantities). Incubate at 35°C in a water bath and observe constantly at intervals of 1 h during the first 4–6 h; if there is no clot formation, observe up to 24 h. Consider a positive result when the clot is completely formed and firm when inverting the tube. On the contrary result, auxiliary tests should be performed such as Gram staining of each bacterial culture, seeking Gram-positive cocci grouped in clusters of grapes, a catalase, and fermentation of glucose and mannitol tests.

Important note: For each new batch of reagents, a coagulation test must be performed on rabbit plasma by adding a drop of 5% calcium chloride to 0.5 mL of reconstituted plasma, forming a clot in 10–15 s.

1.4.1.4. Thermonuclease test

S. aureus produces a thermonuclease enzyme (this capacity is not limited to this species). Detection of staphylococcal thermostable deoxyribonuclease (thermonuclease) in food is used as an indirect test to evident the presence of large amounts of *S. aureus* in food and of staphylococcal enterotoxins.

The production of this enzyme is inhibited by anaerobiosis and is stimulated by the presence of oxygen, it requires calcium ions for its enzymatic activity, its optimal pH is 8.6 and is precipitated with ammonium sulfate. Its thermal stability (resistant to temperatures of 130°C for 16.6 min) is the only association with the growth of *S. aureus*.

When interpreting this test, it is essential to consider the existence of enterotoxigenic strains negatives to both tests. Among the negative coagulase strains, some have developed the ability to synthesize enterotoxins. Although *S. aureus* specie is the typical producer of these toxins, other species exhibit the same behavior: *S. intermedius, S. hyicus, S. warneri, S. epidermidis,* among others [37, 38].

For this test procedure, slides are prepared with 3 mL of toluidine-DNA blue agar. Using a Pasteur pipette makes equally spaced holes in the agar. In a boiling water bath, heat 0.3 mL of bacterial culture in BHI for 15 min. With the use of a Pasteur pipette transfer a drop of bacterial culture to a hole of toluidine blue agar-DNA. Repeat for each strain including the positive and negative controls. Incubate at 35 ± 1°C in a humid chamber from 4 to 24 h. The appearance of a pink halo of at least 1 mm qualifies as a positive test.

See **Table 1** for the characteristics of *S. aureus, S. epidermidis,* and micrococci, in order to identifying the specie isolated or identificated.

Characteristic	*S.aureus*	*S.epidermidis*	Micrococci
Catalase activity	+	+	+
Coagulase production	+	–	–
Thermonuclease production	+	–	–
Lysostaphin sensitivity	+	+	–
Anaerobic utilization of glucose	+	+	–
Mannitol	+	–	–

[a]+, most (90% or more) strains are positive; –, most (90% or more) strains are negative.

Table 1. Typical characteristics of *S. aureus, S. epidermidis,* and micrococci[a].

1.4.2. Molecular methods

The molecular identification is centered of detection and sequencing a specific bacterial DNA and used to identify and classify taxonomically several groups of microorganisms, including bacteria by amplification of specific target region by PCR. Another strategy is amplified by PCR-specific genes that belong to certain species, based on specific features like virulence factors or antibiotics resistance genes [39]. The development of identification techniques for a clinical rapid diagnosis is necessary. The PCR is a rapid, sensitive, and less time-consuming than the conventional bacteriological identification methods [40] and is extensively used to identify bacteria isolated from different kind of samples, including foods [41], soil [42], and infected human tissue [43].

1.4.3. Identification based on 16S ribosomal RNA gene

The 16S rRNA gene is part of all bacteria and is commonly used for taxonomic purposes because it is a highly conserved region; the rate of protection from change is assumed to result from the importance as a serious constituent of cell function [44]. Into the sequence of 16Sr RNA gene are indicated variable regions; Chakravorty et al. [45] in his study describe nine regions with sufficient diversity that are suitable for taxonomic analysis; their investigation determined that V1 hypervariable region best differentiated among *S. aureus* and coagulase-negative *Staphylococcus* sp. Linked to the PCR technique, the use of the 16S rRNA DNA fragment, amplified by using specific oligonucleotides, has been proposed; this gene is widely used for the preparation of phylogenetic trees and useful in finding the evolutionary relationships between two or more individuals [40, 46]. For *S. aureus*, the sequence of 16S ribosomal RNA is reported in the NCBI (National Center for Biotechnology Information) with just over 1.4 kb and access to the GenBank (KP728240.1) was reported by Nazari [47].

1.4.4. *Coa gene*

S. aureus is the only known bacterium that produces coagulase enzyme, which is determined by the *coa* gene. This gene is considered useful and specific for the identification of *S. aureus* and many years ago had been the principal criteria for the separation of *S. aureus* of other *Staphylococcus* that is only ubicuous in this specie [2, 48]. The study of this gene is directly related to the coagulase test. The sequence of *coa* gene is reported in the NCBI (National Center for Biotechnology Information) with just over 2.0 kb and access to the GenBank is AB436964.1 [49].

2. Study case

2.1. Materials and methods

The aim was to determine the incidence of *S. aureus* in fresh lettuce by PCR in order to enhance the efficiency for detection and identification process.

Study zone. Fifty-four lettuce samples were obtained from a company of fresh products located in Sonora, Mexico (**Figure 1**); all the productive process was evaluated, such as cut area, storage, and transportation.

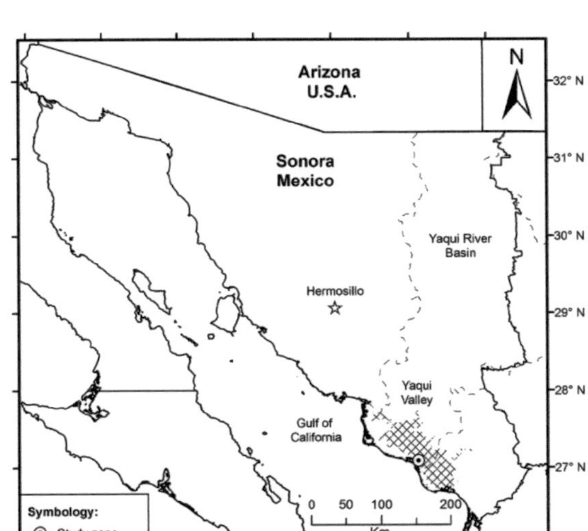

Figure 1. Location of the study area in Sonora, Mexico.

It is important to mention that Mexican vegetable producers strive to be at the top of the market, which involves providing the customers top-quality products all the time. To achieve this goal, they implemented quality control assurance programs as well as partnered with prestigious external certification labs to help they watch every step of the way.

2.2. Transport and handling of sample

Samples were collected as described by Seow et al. [50], briefly, personnel of the laboratory was transferred to the production site in the Yaqui Valley and sampled lettuce from the process, were sampled in original package and immediately placed in sterile resealable bags, and later were transported with iceboxes and stored to 4°C until analysis. Product information such as production date, lote, and "best before" were registered in the database of the study. All the samples were analyzed within 24 h after time of collection, in the meantime, keeping them in their original storage conditions.

2.3. Isolation and identification

The Baird-Parker method was used for isolating pathogens from 54 lettuce samples (**Figure 2**). *S. aureus* ATCC 11632 was used as a positive control; the bacterial strains were cultivated on nutrient agar slant and kept at 4°C. Every 25 days a subculture was carried out to maintain bacterial viability; this process was only repeated two times. A BD Difco™ Tryptic Soy Broth (Soybean-Casein Digest Broth Medium, Ref 211825) was used for the genomic DNA extraction, and the medium was maintained at 30°C with overnight shaking. Genomic DNA extraction was performed according the Mericon DNA Bacteria Plus Kit for Gram-positive bacteria (Qiagen Ref 69534). After extraction procedures, the amount and purity DNA were measured with spectrometer Nanodrop 2000c and the integrity of the DNA on the extracted material in agarose gel electrophoresis was verified.

Figure 2. *S. aureus* in Baird-Parker Agar isolated from lettuce.

The detection by PCR was performed using the pair of primers of *coa* gene [48]. In an Eppendorf PCR tube (0.2 mL), 3 mM $MgCl_2$, 0.8 μM oligonucleotides, 1.6 μM dNTP, 1 unit of Taq DNA polymerase, 10 ng/μl ADN, and buffer 1× were added to make a final volume of 25 μl. A fragment of 674 bp was amplified by the primers *coaf* 5′-ATA GAG CTG ATG GTA CAG G-3′ and *coar* 5′-GCT TCC GAT TGT TCG ATG C-3′), and the PCR protocol was performed in a SimpliAmp Thermal Cycler (Applied Biosystem Ref A24812): 94°C for 10 min (1 cycle), 35 cycles of: 94°C for 30 s, 56°C for 1 min and 72°C for 1 min, with a final cycle of 72°C for 5 min.

Additionally, sensitivity test was performed with base in the genomic DNA concentration. The procedure was done according to Shree et al. [51]; briefly, from genomic DNA of the ATCC 11632 were done dilutions (50, 5, 0.5 ng/μl; 50, 5, 0.5 pg/μl and 50, 5, 0.5 fg/μl). PCR and electrophoresis gel were carried out as we described previously for *coa* gene. The assay was done in triplicates.

The phylogenetic tree was constructed by comparing conserved sequences from the adjacent 16S gene, using the F_2C 5′-AGAGTTTGATCATGGCTC-3′ and C 5′-ACGGGCGGTGTGTAC-3′ primers, in order to obtain a fragment of approximately 1600 bp which was bidirectionally sequenced; the mix reaction was as described above and the PCR conditions were as follows: 95°C for 10 min (1 cycle), 32 cycles of: 95°C for 1 min, 60°C for 1 min and 72°C for 2 min, with a final cycle of 72°C for 5 min.

The purification of PCR products was performed according to the Qiaquick PCR Purification Kit (Qiagen, EUA, Ref 28106), and 400 ng was evaporated in a dry bath at 56°C for 12 h for bidirectional sequencing. The sequences of regions were compared with the National Center of Biotechnology Information (NCBI) data (http://www.ncbi.nlm.gov/) using BLAST-N. The output was grouped such that all members exhibited more than 90% similarity; the alignment of the DNA sequence data was analyzed in Mega 6 software for the phylogenetic tree building with bootstrap analysis (1000 repeats).

To test the antimicrobial effect, the disk diffusion method (Kirby-Bauer) was applied, using Mueller-Hinton agar (MCD Ref 7131) and five antibiotics with different concentrations: tetracycline 30 μg (Oxoid Ref CT0054B), trimethoprim-sulfamethoxazole 25 μg (Oxoid Ref CT0052B), clarithromycin 15 μg (Oxoid Ref CT0693B), oxacillin 1 μg (Oxoid Ref CT0159B), and penicillin G 10 U (Oxoid Ref CT0043B). Each assay was performed in triplicate and the diameter of the inhibition zone was calculated (mm) [52].

3. Results

Figure 3a shows the temperature gradient for *coa* gene, where the range of 56–60°C was observed a specific amplification and 56°C was selected temperature to annealing specific primers in target gene. **Figure 3b** shows the sensitivity of the method for target *coa* gene in range of genomic DNA concentrations (50, 5 and 0.5 ng/μl; 50, 5 and 0.5 pg/μl; 50, 5 and 0.5 fg/μl); the method has a sensitivity up to 0.5 pg/μl.

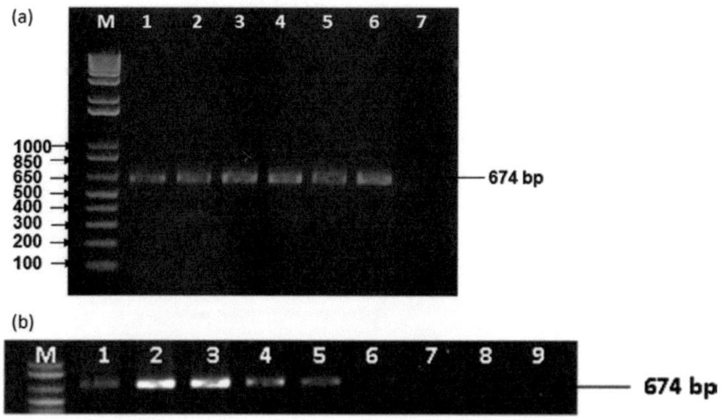

Figure 3. (a) Agarose gel electrophoresis showing temperature gradient for *coa* target gene by PCR with specific primers. M, size marker (1 kb plus DNA Ladder, Invitrogen™); lane 1, 50°C; lane 2, 52°C; lane 3, 54°C; lane 4, 56°C; lane 5, 58°C; lane 6, 60°C and lane 7, negative control (ultrapure water, Invitrogen™). (b) Agarose gel electrophoresis showing the sensitivity of detection for *coa* gene using different DNA concentrations. M, size marker (1 kb plus DNA Ladder, Invitrogen™); lane 1, 50 ng; lane 2, 5 ng; lane 3, 0.5 ng; lane 4, 50 pg; lane 5, 5 pg; lane 6, 0.5 pg; lane 7, 50 fg; lane 8, 5 fg and lane 9, 0.5 fg.

The incidence of *S. aureus* was 1.7% (**Figures 2** and **4**). All the isolates were situated in the ATCC 11632 clade in accordance with other reported sequences belonging to this pathogen in the NCBI database with a bootstrap of 98 (**Figure 5**). Similar reports were given by the GenBank sequences D83357.1, D83355.1, JN315147.1, JN390832.1, JN390831.1, JN315154.1, JN315153.1, JN315151.1, JN315150.1, and JN315149.1 being reported in the same operational taxonomic unit (OTU) with similarities in their morphologic, physiologic, and biochemical characteristics.

All the isolates seemed to be resistant to penicillin G 10 U and were susceptible to oxacillin, tetracycline, clarithromycin, and trimethoprim-sulfamethoxazole (**Figure 6**).

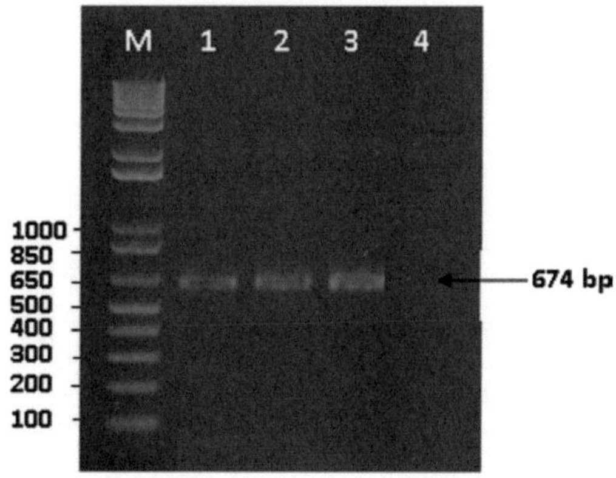

Figure 4. Amplification of target pathogen isolated from lettuce by PCR with specific primers. M, molecular marker (1 kb plus DNA Ladder, Invitrogen™); lane 1, positive control ATCC 11632; line 2, isolated 1; line 3, isolated 2 and line 4, negative control (ultrapure water, Invitrogen™).

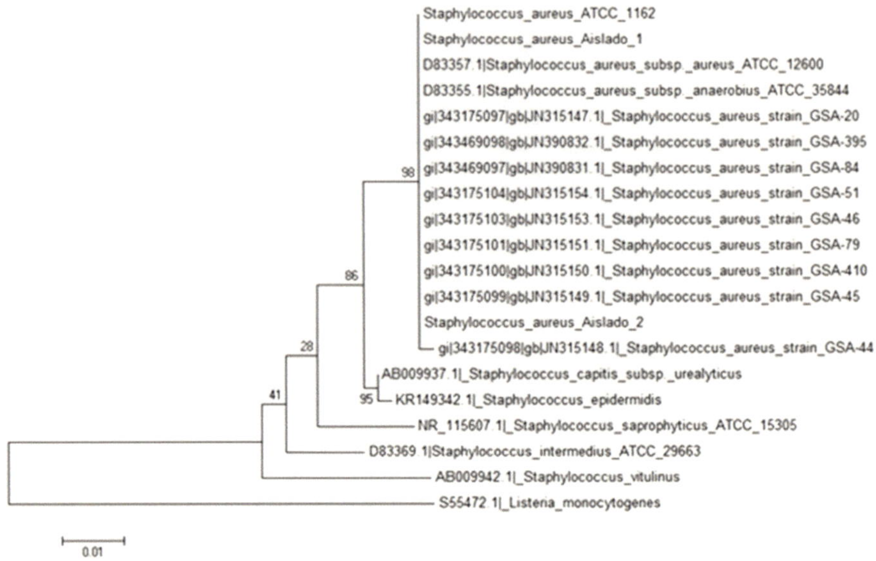

Figure 5. Phylogenetic tree of *S. aureus* maximum verisimilitude constructed from partial sequences of 16S ribosomal DNA of 2 prokaryotic clones and 17 reference sequences obtained from NCBI.

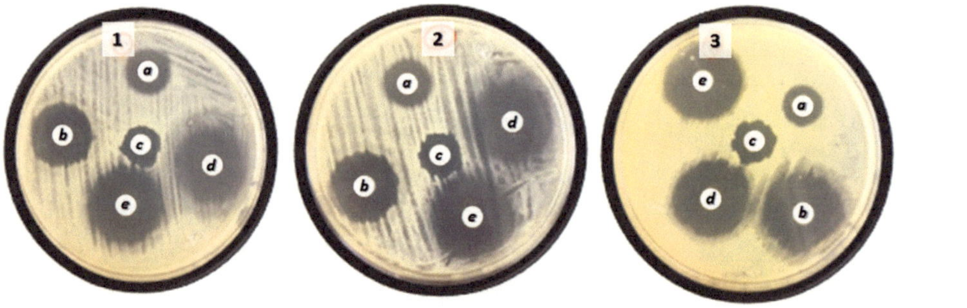

Figure 6. Tests of antibiotic susceptibility; 1: positive control (*Staphylococcus aureus* ATCC11632), 2: isolated 1, 3: isolated 2. (a) Oxacillin 1 μg, (b) trimethoprim-sulfamethoxazole 25 μg, (c) penicillin G 10 U, (d) tetracycline 30 μg, and (e) clarithromycin 15 μg.

4. Discussion

For *coa* gene, the temperature gradient showed that 56°C was the optimal annealing temperature (T_a) for oligonucleotides, showing an adequate specificity for the detection of *S. aureus*. The T_a is defined as the highest temperature where the optimal aligning and amplification occur [53]; this parameter is crucial for the standardization of the method because a low T_a can cause nonspecific amplification, giving undesired PCR products; this is when two or more bands are observed in gel electrophoresis. In this study, the primers features and the correct design lead us to obtain a good and specific amplification in a range of 56–60°C. Likewise, a high T_a can cause a low or non-amplification, reducing the possibility to anneal; for this reason, an optimization of priming temperature is necessary [54]. Additionally, annealing was satisfactory at low DNA concentrations (up to 0.5 pg/μl) showing adequate sensitivity. Isolated from lettuce samples were confirmed by amplification of the 674 bp fragment.

For the strategy with the 16S an optimal annealing temperature of 54°C was established for a fragment of approximately 1400 bp; isolates 1 and 2 were aligned in the same clade as the positive control (ATCC 11632) strain. Clinical animal isolates reported at NCBI D83357.1, D83355.1 and isolated from human throats suffering clinical infections JN315147.1, JN390832.1, JN390831.1, JN315154.1, JN315153.1, JN315151.1, JN315150.1, and JN315149.1 show that isolates 1 and 2 are potentially dangerous if the vegetable is not properly sanitized before consuming.

Low incidence of *S. aureus* is directly related to good manufacturing practice of packing companies, mainly because the exposure time of the product in contact with the exterior is very short. Likewise, the product is never in direct contact with the staff due to the use of hairnets, gloves, face masks, aprons ,and boots, as well as all staff washing and disinfecting their hands before entering work and after toileting.

The bacterial counts found in this study were below the health limit of 10^2–10^3 CFU g^{-1} of *S. aureus* in food set by the Codex Alimentarius, stabilizing a good quality of lettuce with respect to this pathogen. A study by Viswanathan and Kaur [55] reports the presence of *S. aureus* in 23% of a total of 120 samples from various vegetables in India. This incidence is attributed to postharvest and human contamination due to the management of the foods. These results make evident the permanence of the pathogen in this food group, the proper handling of Mexican producers, and the safety of their food.

5. Conclusion

The molecular techniques used in this study are suitable for the identification of *S. aureus* isolated from lettuce, increasing our capability of detecting this pathogen by improving the process and increasing the efficiency, contributing to the safety of this vegetable.

Acknowledgements

AF Chávez-Almanza gratefully acknowledges the Consejo Nacional de Ciencia y Tecnología (CONACYT) by PhD scholarship: 411992. This research was funded by the Instituto Tecnológico de Sonora (Project PROFAPI-00359) and by the Consejo Nacional de Ciencia y Tecnología (PDCPN2014: 248160).

Author details

Andrés F. Chávez-Almanza, Jaime López-Cervantes*, Ernesto U. Cantú-Soto, Dalia I. Sánchez-Machado and Olga N. Campas-Baypoli

*Address all correspondence to: jaime.lopez@itson.edu.mx

Departamento de Biotecnología y Ciencias Alimentarias, Instituto Tecnológico de Sonora, Ciudad Obregón, Sonora, Mexico

References

[1] MacFaddin JF. Pruebas bioquímicas para la identificación de bacterias de importancia clínica, 3er Edición, Ed. Montevideo, Uruguay: Médica Panamericana, 2003, 850 p. ISBN: 950-06-1572-X; 84-7903-789-X.

[2] Torres Vitela MR. Intoxicación estafilocócica: En riesgos asociados al consumo de alimentos Guadalajara, Jalisco, México: Universidad de Guadalajara, 2013, 125–58 p. ISBN: 978-607-450-641-9

[3] Cervantes-García E, García-González R, Salazar-Schettino PM. Características generales del *Staphylococcus aureus*. Rev Latinoam Patol Clin Med Lab. 2014; 61(1): 28–40. ISSN: 04-2013-091711535400-102

[4] Jansen UK, Girgenti DQ, Scully IL, Anderson AS. Vaccine review: *Staphylococcus aureus* vaccines: problems and prospects. Vaccine. 2013; 31: 2723–2730. doi:10.1016/j.vaccine.2013.04.002

[5] Ovchinnikova ES, Van der Mei HC, Krom BP, Busscher HJ. Exchange of adsorbed serum proteins during adhesion of *Staphylococcus aureus* to an abiotic surface and *Candida albicans* hyphae—an AFM study. Colloids Surf B Biointerfaces. 2013; 110: 45–50. doi: 10.1016/j.colsurfb.2013.04.015

[6] Huong BTM, Mahmud ZH, Neogi SB, Kassu A, Nhien NV, Mohammad A, *et al*. Toxigenicity and genetic diversity of *Staphylococcus aureus* isolated from Vietnamese ready-to-eat foods. Food Control. 2010; 21: 166–171. doi:10.1016/j.foodcont.2009.05.001

[7] Lina G, Piemont Y, Godail-Gamot F, Bes M, Peter MO, Gauduchon V, *et al*. Involvement of Panton-Valentine leukocidin-producing *Staphylococcus aureus* in primary skin infections and pneumonia. Clin Infect Dis. 1999; 29(5): 1128–1132. doi:10.1086/313461

[8] Borraz-Ordás M. Epidemiología de la resistencia a meticilina en cepas de "*Staphylococcus aureus*" aisladas en hospitales españoles. [Thesis] Universidad de Barcelona; 2006

[9] Bustos-Martínez JA, Hamdan-Partida A, Gutiérrez-Cárdenas M. *Staphylococcus aureus:* the reemergence of a pathogen in the community. Rev Biomed. 2006; 17(4): 287–305. ISSN 2007-8447

[10] Perazzi B, Camacho M, Bombicino K, Flores Z, Vay C, Famiglietti A. *Staphylococcus aureus:* new and old antimicrobial agents. Rev Arg Microbiol. 2010; 42(3): 199–202. ISSN 1851-7617

[11] Scolari G, Vescovo M, Zacconi C, Bonadé A. Influence of *Lactobacillus plantarum* on *Staphylococcus aureus* growth in a fresh vegetable model system. Eur Food Red Technol. 2004; 218: 274–277. doi:10.1007/s00217-003-0841-x

[12] Fernández E. Microbiología e inocuidad de los alimentos. Universidad Autónoma de Querétaro. Querétaro, México: Segunda edición, 2008. ISBN: 9709263005, 97897092 63008

[13] Ding T, Shin YH, Choi NJ, Ha SD, Chung MS, Hwang IG, *et al*. Mathematical modeling on the growth of *Staphylococcus aureus* in sandwich. Food Sci Biotechnol. 2010; 19: 763–768. doi:10.1007/s10068-010-0107-x

[14] Jeong YS, Jung HK, Hong JH. Multpliex real-time polymerase chain reaction for rapid detection of *Staphylococcus aureus*, *Vibrio parahaemolyticus*, and *Salmonella thypimurium* in milk and kimbap. J Korean Soc Appl Biol Chem. 2013; 56: 715–721. doi 10.1007/s13765-013-3194-6

[15] Zendejas-Manzo GS, Avalos-Flores H, Soto-Padilla MY. Microbiología general de *Staphylococcus aureus*: Characteristics and methods of identifying pathogenicity. Rev Biomed 2014; 25: 129–143. ISSN 2007-8447

[16] Shim WB, Je GS, Kim K, Mtenga AB, Lee WG, Song JU, *et al*. Effect of irradiation on kinetic behavior of *Salmonella Typhimurium* and *Staphylococcus aureus* in lettuce and damage of bacterial cell envelope. Radiat Phys Chem. 2012; 81: 566–571. doi:10.1016/j.radphyschem.2012.01.008

[17] Fetsch A, Contzen M, Hartelt K, Kleiser A, Maassen S, Rau J, *et al*. *Staphylococcus aureus* food-poisoning outbreak associated with the consumption of ice-cream. Int J Food Microbiol. 2014; 187: 1–6. doi:10.1016/j.ijfoodmicro.2014.06.017

[18] Lynch MF, Tauxe RV, Hedberg CW. The growing burden of foodborne outbreaks due to contaminated fresh produce: risks and opportunities [review]. Epidemiol Infect. 2009; 137(3): 307–315. doi:10.1017/S0950268808001969

[19] Callejón R, Rodríguez-Naranjo M, Ubeda C, Hornedo-Ortega R, Garcia-Parrilla M, Troncoso A. Reported foodborne outbreaks due to fresh produce in the United States

and European Union: trends and causes. Foodborne Pathog Dis. 2015; 12(1): 32–38. doi: 10.1089/fpd.2014.1821

[20] Faour-Klingbeil D, Murtada M, Kuri V, Todd ECD. Understanding the routes of contamination of ready to eat vegetables in the Middle East. Food Control. 2015; 62: 125–133. doi:10.1016/j.foodcont.2015.10.024

[21] Lopéz-Galvéz F, Allende A, Truchado P, Martínez-Sánchez A, Tudela JA, Selma MV. Suitability of aqueous chlorine dioxide versus sodium hypochlorite as an effective sanitizer for preserving quality of fresh-cut lettuce while avoiding by-product formation. Postharvest Biol Tech. 2010; 55: 53–60. doi:10.1016/j.postharvbio.2009.08.001

[22] Lee SY, Baek SY. Effect of chemical sanitizer combined with modified atmosphere packaging on inhibiting *Escherichia coli* O157:H7 in commercial spinach. Food Microbiol. 2008; 25: 582–587. doi:10.1016/j.fm.2008.02.003

[23] Wagner M, Auer B, Trittremmel C, Hein I, Schoder D. Survey on the *Listeria* contamination of ready-to-eat food products and household environments in Vienna, Austria. Zoonoses Public Health. 2007; 54(1): 16–22. doi:10.1111/j.1863-2378.2007.00982.x

[24] Christison CA, Lindsay D, Von-Holy A. Microbiological survey of ready-to-eat foods and associated preparation surfaces in retail delicatessens, Johannesburg, South Africa. Food Control. 2008; 19(7), 727–733. doi:10.1016/j.foodcont.2007.07.004

[25] Cho JI, Lee SH, Lim JS, Koh YJ, Kwak HS, Hwang IG. Detection and distribution of food-borne bacteria in ready-to-eat foods in Korea. Food Sci Biotechnol. 2011; 20(2): 525–529. doi:10.1007/s10068-011-0073-y

[26] Cabedo L, Picart i Barrot L, Teixidó i Canelles A. Prevalence of *Listeria monocytogenes* and *Salmonella* in ready-to-eat food in Catalonia, Spain. J Food Prot. 2008; 71(4): 855–859. ISSN 0362-028X

[27] European Commission. Scientific Committee on Food. Risk Profile on the Microbiological Contamination of Fruits and Vegetables Eaten Raw. 2002; (SCF/CS/FMH/SURF/Final)

[28] Buck JW, Walcott RR, Beuchat LR. Recent trends in microbiological safety of fruits and vegetables. Plant Health Prog. 2003; (21). doi:10.1094/PHP-2003-0121-01-RV (http://www.plantmanagementnetwork.org/pub/php/review/2003/safety/)

[29] Harris LJ, Farber JN, Beuchat LR, Parish ME, Suslow TV, Garrett EH, *et al.* Outbreaks associated with fresh produce: incidence, growth, and survival of pathogens in fresh and fresh-cut produce. Compr Rev in Food Sci Food Saf. 2003; 2: 78–141. doi:10.1111/j.1541-4337.2003.tb00031.x

[30] Kisluk G, Hoover DG, Kneil KE, Yaron S. Quantification of low and high levels of *Salmonella enterica* serovar *Typhimurium* on leaves. LWT-Food Sci Technol. 2012; 45(1): 36–42. doi:10.1016/j.lwt.2011.07.029

[31] Parrilla-Cerrillo MC, Vázquez-Castellanos JL, Saldate-Castañeda EO, Nava-Fernández LM. Food-borne toxic infection outbreaks of microbial and parasitic origin. Salud Pública Méx. 1993; 35: 456–463. ISSN: 0036-3634

[32] Schwentesius-Rindermann R, Gómez-Cruz MA. México en el mercado hortícola mundial. Algunos datos. Comercio Exterior. 1994; 341–348

[33] SAGARPA. Info Entidad Anual. [Internet]. 2014. Sicagro.sagarpa.gob.mx. Available from: http://sicagro.sagarpa.gob.mx/SICAGRO_CONSULTA/infoentidadanual.aspx [Accessed 2016-08-12]

[34] Bennett RW, Lancette GA. BAM: *Staphylococcus aureus*. Of FDA U.S. Food and Drug Administration [Internet]. 2001. Available from: http://www.fda.gov/Food/FoodScienceResearch/LaboratoryMethods/ucm071429.htm [Accessed 2016-08-01]

[35] Diario Oficial de la Federación; NORMA Oficial Mexicana NOM-210-SSA1-2014, Productos y servicios. Métodos de prueba microbiológicos. Determinación de microorganismos indicadores. Determinación de microorganismos patógenos. [Internet]. 2015. Available from: http://www.economia-noms.gob.mx/normas/noms/2010/210ssa12015.pdf [Accessed 2016-08-28]

[36] Forbes BA, Sahm DF, Weissfeld AS. Bailey & Scott's Diagnostic Microbiology. Elsevier Mosby, 11th edn., St. Louis MO 2007. ISBN 0808923641, 9780808923640

[37] Valle J, Gómez-Lucia E, Piriz S, Goyache J, Orden JA, Valdillo S. Enterotoxin production by staphylococci isolated from healthy goats. Appl Environ Microbiol. 1990; 56: 1323–1326. ISSN: 1098-5336

[38] Hirooka EY, Mueller EE, Freitos JC. Enterotoxigenicity of Staphylococcus intermedius of canine origin. Int J Food Microbiol. 1988; 7: 185–191. doi:10.1016/0168-1605(88)90036-0

[39] Barghouthi SA. A Universal method for the identification of bacteria based on general PCR primers. Indian J Microbiol. 2011; 51(4):430–444. doi:10.1007/s12088-011-0122-5

[40] Chiang YC, Yang CY, Li C, Ho YC, Lin CK, Tsen HY. Identification of Bacillus spp., *Escherichia coli*, *Salmonella* spp., *Staphylococcus* spp. and *Vibrio* spp. with 16S ribosomal DNA-based oligonucleotide array hybridization. Int J Food Microbiol. 2006; 107: 131–37. doi:10.1016/j.ijfoodmicro.2005.04.028

[41] El-Jakee J, Marouf SA, Ata NS, Abdel-Rahman EH, El-Moez SIA, Samy AA, El-Sayed WE. Rapid method for detection of *Staphylococcus aureus* Enterotoxins in Food. Glob Vet. 2013; 11(3): 335–341. doi:10.5829/idosi.gv.2013.11.3.1140

[42] Figueroa-López AM, Cordero-Ramírez JD, Martínez-Álvarez JC, López-Meyer M, Lizárraga-Sánchez GJ, Félix-Gastélum R, Castro-Martínez C, Maldonado-Mendoza IE. Rhizospheric bacteria of maize with potential for biocontrol of *Fusarium verticillioides*. SpringerPlus. 2016; 5(1): 1–12. doi:10.1186/s40064-016-1780-x

[43] Krimmer V, Merkert H, Von Eiff C, Frosch M, Eulert J, Löhr JF, Ziebuhr W. Detection of *Staphylococcus aureus* and *Staphylococcus epidermidis* in clinical samples by 16S rRNA-directed in situ hybridization. J Clin Microbiol. 1999; 37(8): 2667–2673. ISSN: 1098-660X

[44] Clarridge JE. Impact of 16S rRNA gene sequence analysis for identification of bacteria on clinical microbiology and infectious diseases. Clin Microbiol Rev. 2004; 17(4): 840–862. doi:10.1128/CMR.17.4.840-862.2004

[45] Chakravorty S, Helb D, Burday M, Connell N, Alland D. A detailed analysis of 16S ribosomal RNA genes segments for the diagnosis of pathogenic bacteria. J Microbiol Methods. 2007; 69: 330–339. doi:10.1016/j.mimet.2007.02.005

[46] Lamers RP, Muthukrishnan G, Castoe TA, Tafur S, Cole AM, Parkinson CL. Phylogenetic relationships among *Staphylococcus* species and refinement of cluster groups based on multilocus data. BMC Evol Biol. 2012; 12: 171. doi:10.1186/1471-2148-12-171

[47] NCBI National Center for Biotechnology Information. *Staphylococcus aureus* strain Nazari 16S ribosomal RNA gene, partial sequence Nucleotide—NCBI. [Internet]. 2015. Available from: http://www.ncbi.nlm.nih.gov/nuccore/KP728240.1#feature_901838169 [Accessed 2016-07-27]

[48] Ahmadi M, Rohani SMR, Ayremlou N. Detection of *Staphylococcus aureus* in milk by PCR. Comp Clin Pathol. 2010; 19: 91–94. doi:10.1007/s00580-009-0901-0

[49] Berglund C, Ma X, Ikeda M, Watanabe S, Soderquist B, Ito T. Genetic diversity of methicillin-resistant *Staphylococcus aureus* carrying type IV SCCmec in Orebro County and the western region of Sweden. J Antimicrob Chemother. 2009; 63(1): 32–41. doi: 10.1093/jac/dkn435

[50] Seow J, Ágoston R, Phua L, Yuk HG. Microbiological quality of fresh vegetables and fruits sold in Singapore. Food Control. 2012; 25(1): 39–44. doi:10.1016/j.foodcont.2011.10.017

[51] Shree PT, Ah RH, Jun MC, Jang HH. Multiplex PCR and DNA array for the detection of *Bacillus cereus*, *Staphylococcus aureus*, *Listeria monocytogenes*, *Escherichia coli* O157:H7, and *Salmonella* spp. targeting virulence-related genes. Ann Microbiol. 2012. doi 10.1007/s13213-012-0526-4

[52] Cavalieri SJ, Harbeck RJ, McCarter YS, Ortez JH, Rankin ID. Manual of Antimicrobial Susceptibility Testing. United States of America: University of Washington. 2005. ISBN 1-55581-349-6

[53] Wu DY, Ugozzoli L, Pal BK, Qian J, Wallace RB. The effect of temperature and oligonucleotide primer length on the specificity and efficiency of amplification by the polymerase chain reaction. DNA Cell Biol. 1991; 10(3): 233–238. doi:10.1089/dna.1991.10.233

[54] Rychlik W, Spencer WJ, Rhoads RE. Optimization of the annealing temperature for DNA amplification in vitro. Nucleic Acids Res. 1990; 18(21): 6409–6412. doi:10.1093/nar/18.21.6409

[55] Viswanathan P, Kaur R. Prevalence and growth of pathogens on salad vegetables, fruits and sprouts. Int J Hyg Environ Health. 2001; 203: 205–213. doi:10.1078/S1438-4639(04)70030-9

Section 5

Prevention and Treatment of *Staphylococcus aureus*

Chapter 8

Prevention of *Staphylococcus aureus* Contamination on Animal Products Using Indonesian Natural Products

Irma Isnafia Arief

Additional information is available at the end of the chapter

http://dx.doi.org/10.5772/66045

Abstract

Foodborne transmission of pathogenic microorganisms has been recognized as an important hazard. One of foodborne pathogens that was well known for 30 years, that associated with animals, have presented as illness-causing agents in humans, is *Staphylococcus aureus*. *S. aureus* is a bacterium that produces enterotoxin, causing poisoning to humans. These bacteria are found in foods that contain high protein such as sausage, eggs, meat, beef, poultry products, and milk products. *S. aureus* is a Gram-positive bacterium that is an indicator of contamination from the worker and tools. *S. aureus* contamination on raw animal products such as eggs, raw beef, and poultry products also milks in Indonesia has been reported by many researchers. Indonesia is a tropical country that has high humidity, heavy rain, and two seasons (dry and wet) that contribute to *S. aureus* contamination especially in animal products. Furthermore, poor postmortem handling on animal products also causes the contamination. Preventive methods are needed for food processing and food storage especially for animal products in Indonesia. This chapter in this book explains the contamination of *S. aureus* in animal products in Indonesia and the preventive methods used in Indonesia to reduce the contamination. Plant extracts, herbs, spices, bacteriocins, and lactic acid bacteria have been widely used in food processing in Indonesia that proved as biopreservatives for animal products.

Keywords: prevention, *Staphylococcus aureus*, animal products, Indonesia

1. Introduction

Foodborne transmission of pathogenic microorganisms has been recognized as an important hazard. The predominant foodborne pathogens that were known 30 years ago are *Salmonella, Clostridium botulinum, Clostridium perfringens, and Staphylococcus aureus* and have been joined by a widening array of pathogens of bacterial, viral, and parasitic origin. Those pathogens that were only seen associated with animals have been presented as illness-causing agents in humans [1]. *S. aureus* is a Gram-positive bacterium that is an indicator of contamination from the workers and tools. *S. aureus* is a normal flora on the skin and in the respiratory organs in humans, and it is generally found in 20–50% of healthy population [2]. *S. aureus* contamination in food could also occur after the food has been cooked. In relation to cases of food poisoning, *S. aureus* enterotoxin intoxication on consumers occurs through the establishment of contamination on food consumed. This enterotoxin is resistant to heat (heat stable), acid-resistant, and resistant to the effects of proteolytic enzymes such as pepsin and trypsin.

S aureus contamination is found in animal products that are marketed in Indonesia, such as eggs, chickens, and raw beef and also other raw meat products. Poor handling and improper storage methods cause the contamination of *S. aureus,* which survive on kitchen utensils and unwashed hands.

Some *S. aureus* contamination that is in animal products marketed in Indonesia is as follows:

a. Chicken eggs

Egg contamination can be derived from the environment. *S. aureus* would stick to the eggshell and subsequently on holding it penetrates into the egg through the pores in the eggshell.

b. Chicken meats

Population of *S. aureus* contamination on chicken breast meats is 2.71 ± 0.02 log CFU/g up to 3.25 ± 0.28 log CFU/g in Java Island, Indonesia (research result). *S. aureus* contamination on breast chicken meat can be derived from the contents of the digestive tract during slaughtering process in the poultry abattoir. Contamination on the carcass also occurs from the air or feces that contaminates skin and carcass [3]. External factors that influence the contamination of *S. aureus* are pH value and a_w (water activity) on breast chicken meat. a_w that is optimum in food for growth factor of *S. aureus* is 0.8–1.0.

c. Beef

Contamination of *S. aureus* on fresh beef at traditional market in West Java, Indonesia, has been investigated. The population of *S. aureus* was approximately 2.48 log CFU/g [4] and increased continuously every hour in the room temperature of storage.

2. Antimicrobials agents against *Staphylococcus aureus*

2.1. Spices, herbs and plant extract as antimicrobials against *Staphylococcus aureus*

The compounds found in herbs and found beneficial as traditional medicine can also be used as antibacterials and natural preservatives. The use of antibacterial synthetic or synthetic preservatives in foods such as the addition of formaldehyde or borax (borax) if taken continuously will cause a disease. The existence of the above phenomenon encourages people to find the best solution for health. An alternative solution is to replace synthetic antibacterial agents with natural antibacterial agents. Preventive methods have been used through the application of indigenous Indonesian herbs, plant extracts, and spices. The antimicrobial activities of plant extracts used for seasoning in foods have been recognized. The most common plant secondary metabolites that have antimicrobial activities occur in the following groups: alkaloids, anthraquinones, coumarins, essential oils (terpenoid and phenylpropanoids), flavonoids, steroids, and triterpenoids [5]. Some of them have antimicrobial activities.

2.1.1. *Curcuma domestica val*

Turmeric (*Curcuma domestica* val) is one of the plants that is used for traditional medicine by our ancestors long ago. Turmeric has great potential in the pharmacological activity that is, anti-inflammatory, anti-immunodeficiency, anti-virus (bird flu virus), anti-bacterial, and anti-fungal [6]. The antibacterial properties in turmeric are caused by the chemical content of its main and essential oil curcuminoid.

2.1.2. *Ginger*

Ginger can grow in the lowlands of the mountainous regions with an altitude of 0–1500 m above sea level. It has been used in food for seasoning in Indonesia. Meat cooked with ginger can have longer storage duration than without ginger. Ginger contains gingerol bioactive compound, which is a major component that can be converted into shogaol or zingerone shogaol formed from gingerol during the heating process [7].

2.1.3. *Garlic*

Raw garlic can be minced, pressed, sautéed, pickled, boiled, and juiced. Garlic sulphur compound(s) is(are) the primary bioactive agent(s). The major thiosulfanates, allicin, account for approximately half of the total of thiosulfanates from the *Allium sativum* genus [8]. Allicin was described as colorless oil, extremely pungent for the principal odor and taste of garlic. It was reported that allicin in concentrations of 1: 85,000 in broth was bactericidal to a wide variety of Gram-negative and Gram-positive organisms. A 5% garlic extract concentration has a germicidal effect on *S. aureus* [9]. Garlic extract used for seasoning in Indonesian food has strong antibacterial activities against *S. aureus* (**Figure 1**).

2.1.4. *Clove oil*

Clove oil has a potential as a preservative for food products and is known as Generally Recognized As Safe (GRAS) as a food ingredient. In addition, various studies have shown that clove oil has antimicrobial properties against *Salmonella* sp., *Listeria monocytogenes, Pseudomonas aeruginosa, Bacillus subtilis, Escherichia coli,* and *S. aureus*. An amount of 0.25/100 ml of clove oil could inhibit *S. aureus*. The application of clove oil in processed meat products showed that at a concentration of 1 ml/l, it reduces the bacterial population significantly ($P < 0.05$), as much as 0.88 log CFU/g.

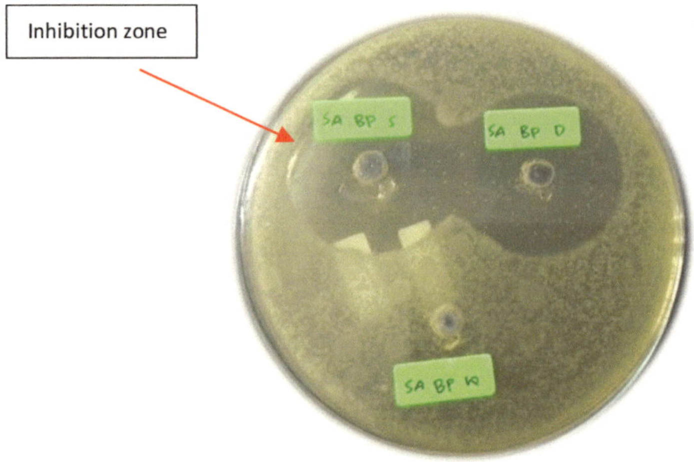

Figure 1. Inhibition zone of antimicrobial activities of garlic extract against *S. aureus*.

2.1.5. Roselle flower

An essential ingredient contained in rosella flower petals is the pigment anthocyanin that forms flavonoids and acts as antioxidants. Anthocyanin that causes the red color of this plant contains delfinidin-3-siloglukosida, delfinidin-3-glucoside, and sianidin-3-siloglukosida, while flavonoids contain gossypetin and mucilage (rhamnogalacturonan, arabinogalactan, and arabinan). *Hibiscus sabdariffa* Linn (Roselle flower) also contains phenol compounds that can be chemically defined by the presence of the aromatic ring carrying one (phenol) or more (polyphenols) substitution of hydroxyls [10]. The working of phenol in killing microorganisms is by cell protein denaturation. Phenol derivatives interact with bacterial cells through adsorption process involving hydrogen bond. At low levels, protein complex forms phenol by weak bonds and immediately occurs as decomposition, followed by phenol penetration into cells, causing precipitation and protein denaturation. Roselle flower extracts were proven for their antibacterial activities against *S. aureus* (**Figure 2**) and are used in yoghurt products in Indonesia as flavoring and preservatives.

2.1.6. Red dragon fruit extract

Flavonoids, phenols, hydroquinones, and saponins are the phytochemical compounds found in red dragon fruit peel extract. Steroids and triterpenoids compounds are also found in the red dragon fruit peel. The phytochemical substances of red dragon fruit extract have antibacterial activity that reacts with the bacterial cell wall proteins.

Red dragon fruit peels were extracted by modification maceration. Dragon fruits were cleaned and peeled manually before being cut into small sizes (2 mm). Red dragon fruit peels were dried at 50°C with an oven and ground to a powder. Peel powder was added with a solvent (1:50) for 60 min and filtered. The solution was evaporated at a vacuum evaporator temperature of 60°C. The extract was stored at −20°C and continued to be used in the

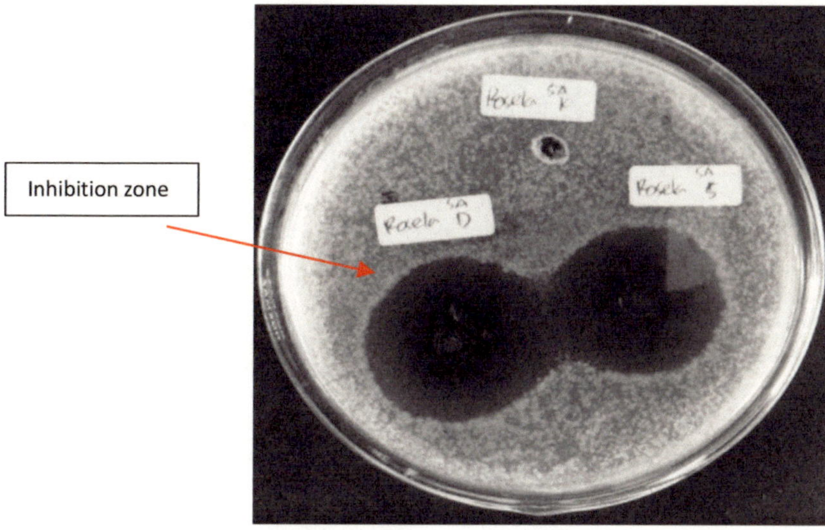

Figure 2. Inhibition zone of antimicrobial activities of Roselle flower extract against *S. aureus*.

antimicrobial analysis. Analysis of antimicrobial activity was performed by the well diffusion method. Bacterial culture was inoculated in NaCl 0.85% to obtain the bacteria concentration of 10^8 CFU mL^{-1}. The dilution of the bacterial culture was done to obtain a culture concentration of 10^6 CFU mL^{-1}. The other culture was grown in Mueller-Hinton Agar medium (Difco™, USA) and provided with holes as well with a predetermined diameter. Extracts were inserted into the well and covered with filter paper. Grail was stored in a refrigerator for 2–3 h, followed by incubation at 37°C for 24 and 48 h. The antimicrobial activity was characterized by the formation of clear zones around wells and measured for its diameter (mm). The inhibition zone produced by red dragon fruit peel extracts showed strong antibacterial activity (**Figure 3**).

Gram-positive bacterium, *S. aureus* ATCC 25923, was more sensitive to the antibacterial activity of red dragon fruit peel extract. The Gram-positive bacteria are more susceptible to antibacterial activity due to the absence of a lipoprotein wall that is capable of preventing antimicrobial compounds. Red dragon fruit peel extract due to its antibacterial compounds such as phenolic compounds could inhibit the growth of bacteria [13]. Application of red dragon fruit extract on beef sausages showed that *S. aureus* was not detected during 20 days of cold storage.

2.1.7. Teak leaf extract

Teak leaf extracts have a composition of flavonoids, alkaloids, tannins, anthraquinones, and naphthoquinones as antimicrobial substances that inhibit the growth of bacteria [11]. Addition of teak leaf extracts effectively inhibited *S. aureus* in the sausages. The 0.5 and 1% concentrations of teak leaf extracts addition on sausage formula in the processing could effectively inhibit *S. aureus* [12]. The method of teak leaf extraction is as follows: The extraction of teak

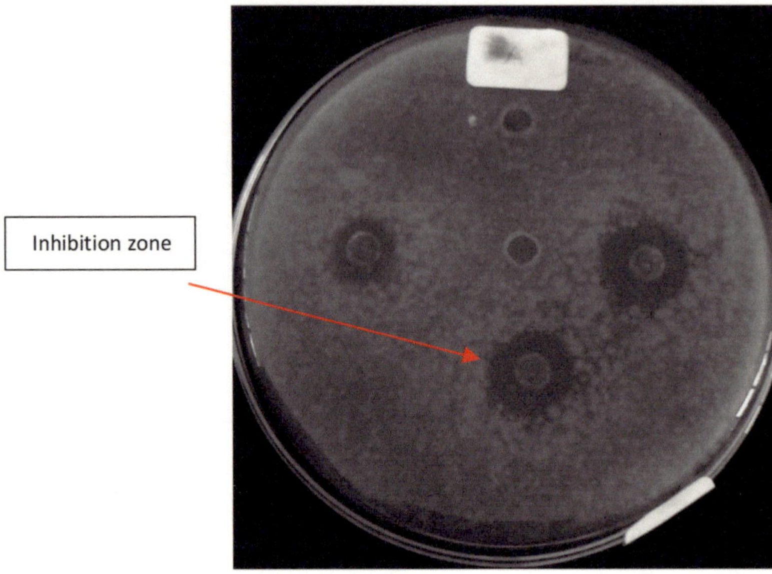

Figure 3. Inhibition zone of antimicrobial activities of red dragon fruit extract against *S. aureus*.

leaf was performed using ethanol extraction. Fresh teak leaf was oven-dried at 60°C for 24 h, chopped, and blended. Two hundred milliliters of 96% ethanol was then added into 20 g of teak leaf powder (10:1 ratio) and was boiled using waterbath at 70°C for 2 h. The mixture was centrifuged at 6000 rpm for 15 min. Ethanol was then removed by air-dry evaporation. The inhibition zone of antimicrobial activities was performed using diffusion methods [12]. The result of the antibacterial activities of teak leaf extract against *S. aureus* is shown in **Figure 4.**

2.2. Bacteriocins as antimicrobials and their application as meat product biopreservatives

Bacteriocins produced by Indonesian lactic acid bacteria *Lactobacillus plantarum* IIA-1A5 was purified and characterized. Plantaricin IIA-1A5 has been previously isolated from Indonesian lactic acid bacteria of *L. plantarum* IIA-1A5. This plantaricin has been shown

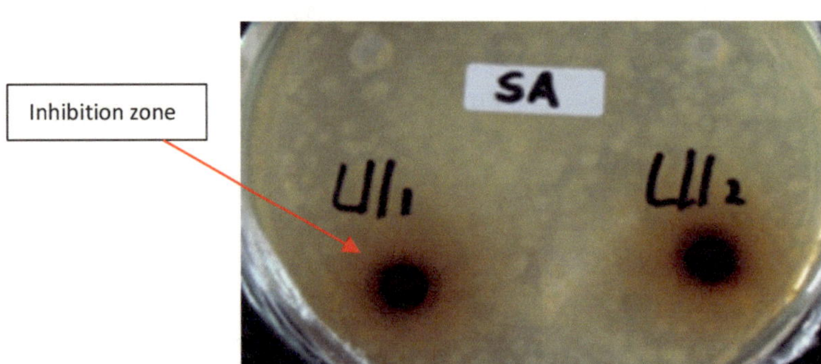

Figure 4. Inhibition zone of antimicrobial activities of teak leaf extract against *S. aureus*.

to inhibit the growth of *S. aureus* [13], making it a promising preservative substance to replace the use of chemical preservatives. Plantaricin could be digested by trypsin enzyme. It was heat stable at 80°C for 30 min and 121°C at 15 min, also active in a broad pH range of 4.0–9.0. Plantaricin IIA-1A5 could inhibit the growth of pathogenic bacteria, such as *E. coli, Salmonella Typhimurium, Bacillus cereus,* and *S. aureus.* Plantaricin IIA-1A5 showed good characteristics as an antimicrobial [14]. Plantaricin IIA-1A5 employs bactericidal activity since it disrupts the cell membrane and promotes the release of ions, proteinaceous, and genetic materials [13]. The cell wall of Gram positive has a thicker peptidoglycan layer, which is dominantly composed of lipoteichoic acid (LTA). The LTA is the target recognition of bacteriocin, facilitating the absorption of bacteriocin in the cell wall of Gram-positive bacteria [15].

2.2.1. Genes involved in the production of plantaricin

The genes responsible for bacteriocin production in *L. plantarum* IIA-1A5 are at least organized in two different operons: *plnABCD* and *plnEFI* [13]. The genes have been sequenced. PlnB (representative of operon plnABCD) amino acid sequence is derived from translation of partial DNA sequences using the software APE plasmid editor (http://biologylabs.utah.edu/jorgensen/wayned/ape/). Regardless of its open-frame reading, partial sequence of plnB is shown in **Figure 5**.

To identify what kind of protein is encoded by plnB gene, we performed protein BLAST. BLAST results showed that the DNA sequence has 100% similarity with the histidine kinase genes for plantaricin on *L. plantarum* (**Figure 6**). The histidine kinase has been reported as one of the genes responsible for bacteriocin production. It is located in the locus responsible for plantaricin production in some of the plantarum strain. Histidine kinase is a quorum sensor to monitor the cell density of a bacterial population. At a certain concentration threshold, histidine kinase will be activated through a certain mechanism and induced with the production of bacteriocin [13].

Sequencing of plnEF gene and translation plnEF gene to amino acid has been conducted. PlnEF amino acid sequences were also obtained from the translation of DNA sequences using the software ApE plasmid editor (http://biologylabs.utah.edu/jorgensen/wayned/ape/). PlnB translation of DNA sequences to amino acid sequences is presented in **Figure 7**.

```
  1 TTC AGA GCA AGC CTA AAT GAC GGT AGT ATT GCA AGT GTT CAA CAT TTG AAG AAT GAG ATA  60
  1 Phe Arg Ala Ser Leu Asn Asp Gly Ser Ile Ala Ser Val Gln His Leu Lys Asn Glu Ile  20

 61 TTG CGC GGG TTA GTT GTA CAG AAG TTT TTT TAT GCG AAA CAG TGT GGG GTT AAG TTG ACG  120
 21 Leu Arg Gly Leu Val Val Gln Lys Phe Phe Tyr Ala Lys Gln Cys Gly Val Lys Leu Thr  40

121 ATT GAA ATA GCT AAC ACT GAC TTT ATT CTC AGT CAT GGT GTT ACA GTG  168
 41 Ile Glu Ile Ala Asn Thr Asp Phe Ile Leu Ser His Gly Val Thr Val  56
```

Figure 5. Amino acid sequences translation of plnB derived from its DNA sequences. The number on the left and right side shows the numbering sequence of the DNA sequence (top row) and amino acids (the second row). Bases in DNA are written in capital letters, while the amino acids are written in the format of three letters.

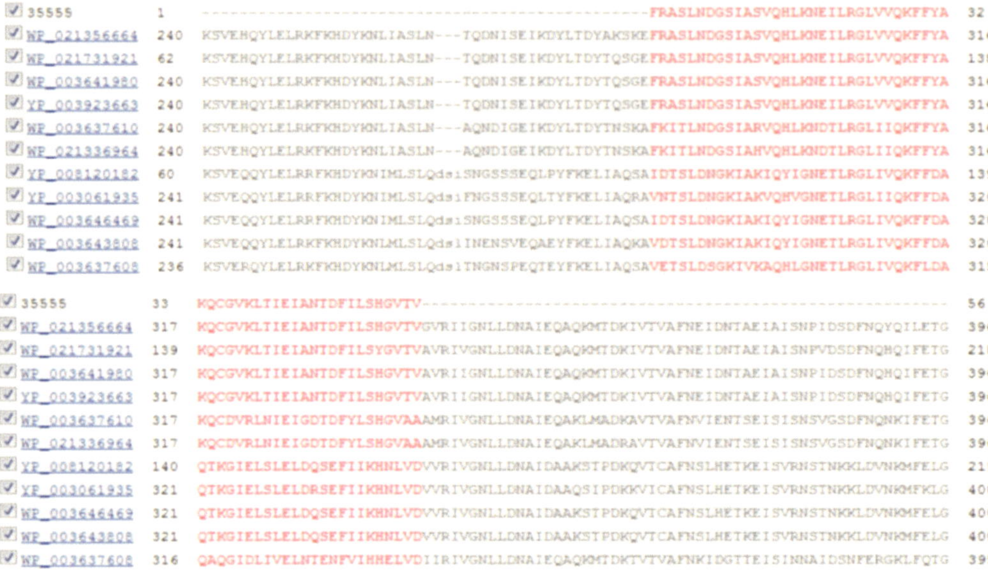

Figure 6. Multiple amino acid sequence alignment PlnB (35555) with homologous proteins. Each homologous proteins used in the alignment presented in the access code in the database. The red sequence shows the location of homology in the alignment.

```
  1 GGCATAGTTTAAATTCCCCCCATATAAACTAAAAAGGCTGAGCACTGCAGGCCCAGTCAT  60
  1  G  I  V  *  I  P  P  I  *  T  K  K  A  E  H  C  R  P  S  H   20

 61 AAATCGAGTGCTTTCTTATAGTGCTTAAACTTGATGGCTTGAAGAACTATCCGTGGATGA 120
 21  K  S  S  A  F  L  *  C  L  N  L  M  A  *  R  T  I  R  G  *   40

121 ATCCTCGGACAGCGCTAATGACCCAATCGGCAGGCCCAACAGCACTTTTATAATTATTCC 180
 41  I  L  G  Q  R  *  *  P  N  R  Q  A  Q  Q  H  F  Y  N  Y  S   60

181 GAACGCCACGCGCGCTATAGGCATGGAAAACGCCACCTGAAATAGCATTTAATTCACGGT 240
 61  E  R  H  A  R  Y  R  H  G  K  R  H  L  K  *  H  L  I  H  G   80

241 CACGCAAAACTAGAAATTTTTTCATAATTGTTGATCTCCCCCAAGAAAATTAACGAATAC 300
 81  H  A  K  L  E  I  F  S  *  L  L  I  S  P  K  K  I  N  E  Y  100

301 TTTTCAAAATACCACGAATGCCTGCAACTGAACCAATTGCATCAACAACATGTCGAACAC 360
101  F  S  K  Y  H  E  C  L  Q  L  N  Q  L  H  Q  Q  H  V  E  H  120

361 TTTTACCAAAGTTATAACCGCCCCGATTAAAACCACCAGATATTTTGGCAAGCTTTTTTT 420
121  F  Y  Q  S  Y  N  R  P  D  *  N  H  Q  I  F  W  Q  A  F  F  140

421 GCGGGCAACCTG 432
141  A  G  N  L  144
```

Figure 7. PlnE translation of DNA sequences into amino acids. The number on the left and right side shows the numbering sequence of the DNA sequence (top row) and amino acids (the second row). Bases in DNA are written in capital letters, while the amino acids are written in one letter.

BLAST results showed that the DNA sequence of plasmid Ape editors has a high homology with the pln locus from several strains of plantaricin from *L. plantarum*. This means plnE correct encoding plantaricin [13]. Alignment results either in whole or in part show that the homology of plnE is more than 90% with various strains of the plantaricin (**Figure 8**).

http://dx.doi.org/10.5772/66045

```
pln5-3          GIVKIPPITKKAEHCRPSHKSSAFLCLNLMA-TIRGILGQRPNRQAQQHFYNYSERHARY 59
BFE5092         GIVKIPPITKKAEHCRPSHKSSAFLCLNLMA-TIRGILGQRPNRQAQQHFYNYSERHARY 59
plnC11          GIVKIPPITKKAEHCRPSHKSSAFLCLNLMA-TIRGILGQRPNRQAQQHFYNYSERHARY 59
pln8p-3         GIVKIPPITKKAEHCRPSHKSSAFLCLNLMA-TIRGILGQRPNRQAQQHFYNYSERHARY 59
plnYM4-3        -IVKIPPITKKAEHCRPSHKSSAFLCLNLMA-TIRGILGQRPNRQAQQHFYNYSERHARY 58
plnEF IIA-1A5   GIV-IPPITKKAEHCRPSHKSSAFLCLNLMARTIRGILGQRPNRQAQQHFYNYSERHARY 59
                 ** *************************** **************************

pln5-3          RHGKRHLKHLIHGHAKLEIFSLLISPKKINEYFSKYHECLQLNQLHQQHVEHFYQSYNRP 119
BFE5092         RHGKRHLKHLIHGHAKLEIFSLLISPKKINEYFSKYHECLQLNQLHQQHVEHFYQSYNRP 119
plnC11          RHGKRHLKHLIHGHAKLEIFSLLISPKKINEYFSKYHECLQLNQLHQQHVEHFYQSYNRP 119
pln8p-3         RHGKRHLKHLIHGHAKLEIFSLLISPKKINEYFSKYHECLQLNQLHQQHVEHFYQSYNRP 119
plnYM4-3        RHGKRHLKHLIHGHAKLEIFSLLISPKKINEYFSKYHECLQLNQLHQQHVEHFYQSYNRP 118
plnEF IIA-1A5   RHGKRHLKHLIHGHAKLEIFSLLISPKKINEYFSKYHECLQLNQLHQQHVEHFYQSYNRP 119
                ************************************************************

pln5-3          DNHQIFWQAFFAAT- 133
BFE5092         DNHQIFWQAFFAAT- 133
plnC11          DNHQIFWQAFFAAT- 133
pln8p-3         DNHQIFWQAFFAAT- 133
plnYM4-3        DNHQIFWQAFFAAT- 132
plnEF IIA-1A5   DNHQIFWQAFFAGNL 134
************..
```

Figure 8. Multiple sequence alignment of amino acid sequencing of plantaricin EF.

2.2.2. Application of plantaricin IIA-1A5 as a biopreservative

Plantaricin IIA-1A5 could be used as biopreservatives for raw beef after slaughtering from abattoir. The initial contamination of *S. aureus* on raw beef is 2 log CFU/g. The addition of 0.2% plantaricin IIA-1A5 by spraying it onto raw beef surface could enhance the safety of beef from *S. aureus* contamination. Population of *S. aureus* on beef with 0.2% plantaricin is lower than maximum standard allowed by Indonesian standard of fresh beef (2 log CFU/g). In control (without plantaricin addition), population of *S. aureus* increased continuously every hour (3 log CFU/g). Plantaricin IIA-1A5 is able to extend the shelf life of meat stored at room temperature, according to physicochemical and microbiology quality [4].

Another application of plantaricin is as a biopreservative in meat products. *S. aureus* has been observed in meatballs without preservatives after 5 h of storage at room temperature. The 0.3% plantaricin IIA-1A5 addition displayed inhibition of *S. aureus* to be as strong as 0.3% nitrite. Until 20 h storage at room temperature, meatballs with nitrite or plantaricin IIA-1A5 were considerably safe to be consumed, which is a proven and promising potential use of plantaricin as a nitrite replacer for meatballs preservative [16].

3. Bacteriocin produced by *S. aureus*

S. aureus produced bacteriocins and bacteriocin-like substances that were correlated with the presence of a plasmid usually involved in type B exfoliative toxin production. The bacteriocinogenic plasmids carried by the *S. aureus* strains are identified as plasmids larger than 40 kb that code for a high-M bacteriocin and that do not confer immunity [17]. *S. aureus* was isolated from bovine mastitis cases in 56 different Brazilian dairy herds and has been successfully investigated to produce antimicrobial substance (AMS). The bacteriocins may possess potential practical applications since they were able to inhibit important pathogens such as *B. cereus* and *L. monocytogenes* isolated from nosocomial infections [18] and show a potential application in food preservation [19]; meanwhile, the pathogenicity of *S. aureus* should be discussed for safety of bacteriocin. The antimicrobial activity of the bacteriocin produced by *S. aureus* is detected to be resistant to heat treatment at 65°C; however, treatment at 80°C completely abolished its antimicrobial properties [19].

Although *S. aureus* also produced bacteriocin, it could not kill and inhibit the cell itself because of immunity system. Bacteriocin-producing bacteria protect themselves from similar bacteriocin by immunity proteins. When these proteins are expressed in sensitive cells, they strongly protect against externally added similar bacteriocin. The immune system can work synergistically to protect the producing cells from their own bacteriocin [17]. Plasmid carried by the *S. aureus* strains confers immunity identified as small plasmids (8.0–10.4 kb), which code for bacteriocins or bacteriocin-like substances with a low M [18].

4. Conclusion

To prevent *S. aureus* contamination, many antimicrobial substances originated from Indonesia and are widely used for food processing and as preservatives. Herbs, plant extracts, spices, and indigenous bacteriocin isolated from Indonesian lactic acid bacteria also prove to be effective antimicrobial agents for animal products. Many different types of mode of action and antimicrobial mechanisms could be synergic as animal product preservatives in Indonesia.

Author details

Irma Isnafia Arief

Address all correspondence to: irma_isnafia@yahoo.com; irmaisnafia@gmail.com

Department of Animal Production and Technology, Faculty of Animal Science, Bogor Agricultural University (IPB), Bogor, West Java, Indonesia

References

[1] Carry JW, Linz JE, Bathnagar D. Microbial Foodborne Diseases: Mechanisms of Pathogenesis and Toxin Synthesis. Technomic Publishing Company, Inc., Lancaster, Pennsylvania, USA. 2000.

[2] Teixeira P, Silva S, Aranjo F, Azeredo J, Oliveira. Bacterial adhesion to food contacting surfaces, p. 8. Downloaded from https://repositorium.sdum.uminho.pt/bitstream/1822/7711/1/Pilar. 2007

[3] Berrang M, Windham W, Meinersmann R. *Campylobacter, Salmonella,* and *Escherichia coli* on broiler carcasses subjected to a high pH scald and low pH post pick chlorine dip. Poultry Science. 2011; 90(4): 896–900.

[4] Sihombing DE, Arief II, Budiarti S. Application of antimicrobial agents produced by *Lactobacillus plantarum* IIA-1A5 as natural preservative on beef during room temperature storage. Advance Journal of Food Science and Technology. 2015; 8: 251–255.

[5] Cannell, RJP. Natural Products Isolation: Methods in Biotechnology. Humana Press, Totowa, New Jersey. 2003.

[6] Joe B, Vijaykumar M, Lokesh BR. Biological properties of curcumin-cellular and molecular mechanisms of action. Critical Review in Food Science and Nutrition. 2004; 44(2): 97–112.

[7] Wohlmuth H, Smith MK, Brooks LO, Myer SP, Leach DN. Essential oil composition of diploid and tetraploid clones of ginger (Zingiber officinale Roscose) grown in Australia. J.Agric.Food Chem. 2006; 54: 1414–1419.

[8] Bidlack WR, Omaye ST, Meskin MS, Topham DKW. Phytochemicals as Bioactive Agents. Technomic Publishing Co. Inc., Lancaster-Basel. 2000.

[9] Davidson PM, Branen AL. Antimicrobials in Foods. Second edition. Marcell Dekker, Inc., New York. 1993.

[10] Duke JA. Handbook of Energy Crops: Hibiscus Sabdariffa Linn, Malvacea, Roselle. Center for New Crops and Plants Products. Purdue University, West Lafayette, Indiana. 1998.

[11] Purushotham KG, Arun P, Jayarani JJ, Vasnthakumari R, Sankar L, Reddy BR. Synergistic in vitro antibacterial activity of *Tectona grandis* leaves with tetracycline. International Journal of Pharm Tech Research. 2010; 2(1): 519–523.

[12] Arief II, Suryati T, Afiyah DN, Wardhani DP. Physicochemical and organoleptic of beef sausages with teak leaf extract (*Tectona grandis*) addition as preservative and natural dye. International Food Research Journal. 2014; 21: 2033–2042.

[13] Arief II, Budiman C, Jenie BSL, Andreas E, Yuneni A, Plantaricin IIA-1A5 from *Lactobacillus plantarum* IIA-1A5 displays bactericidal activity against *Staphylococcus aureus*. Beneficial Microbes. 2015; 6: 603–613.

[14] Arief II, Jakaria, Suryati T, Wulandari Z, Andreas E. Isolation and characterization of plantaricin produced by *Lactobacillus plantarum* strains (IIA-1A5, IIA-1B1, IIA-2B2). Media Peternakan (Journal of Animal Science and Technology). 2013; 36: 91–100.

[15] Kjus T, Lutzow-Holm C, Christensen OB. Treatment of panniculitis associated with alpha-1-antitrypsin deficiency with alpha-1-protease inhibitor. Acta Dermato-Venereologica, 2002; 83: 462–463.

[16] Kia KW, Arief II, Sumantri C, Budiman C. Plantaricin IIA-1A5 from *Lactobacillus plantarum* IIA-1A5 retards pathogenic bacteria in beef meatball stored at room temperature. American Journal of Food Technology 2016; 11(1–2): 37–43.

[17] Bharti V, Mehta A, Singh S, Jain N, Ahirwal L, Mehta S. Bacteriocin: a novel approach for preservation of food. International Journal of Pharmacy and Pharmaceutical Sciences. 2015; 7: 20–29.

[18] Ceotto H, Nascimento Jdos S, Brito Pdo MAV, Bastos Mdo C. Bacteriocin production by *Staphylococcus aureus* involved in bovine mastitis in Brazil. Research Microbiology. 2009; 160(8): 592–599.

[19] Ceotto H, Brede D, Salehian Z, Nascimento Jdos S, Fagundes PC, Nes IF, Bastos Mdo C. Aureocins 4185, bacteriocins produced by *Staphylococcus aureus* 4185: potential application in food preservation. Foodborne Pathogens and Disease. 2010; 7(10): 1255–1262. doi: 10.1089/fpd.2010.0578.

Chapter 9

Bacteriophage Therapy: An Alternative for the Treatment of *Staphylococcus aureus* Infections in Animals and Animal Models

Claudia I. Barrera-Rivas, Norma A. Valle-Hurtado,
Graciela M. González-Lugo,
Víctor M. Baizabal-Aguirre, Alejandro Bravo-Patiño,
Marcos Cajero-Juárez and Juan J. Valdez-Alarcón

Additional information is available at the end of the chapter

http://dx.doi.org/10.5772/65761

Abstract

Staphylococcus aureus causes hospital-acquired (HA), community-acquired (CA) and companion animal and livestock-associated (LA) infections. Molecular epidemiology studies suggest that although host specificity may be associated with specific genetic lineages, recent human-to-animal and animal-to-human transmissions related to mobile genetic elements have been described. Gene transfers include virulence and antibiotic resistance genes, thus making it difficult to control multidrug resistance *S. aureus* infections. Bacteriophages (phages) and endolysins, the enzymes responsible for bacterial lysis by phages, are alternatives to the use of antibiotics for the control of *S. aureus* infections. In this work, we review current advances in the development of phage therapy and the study and design of recombinant endolysins to treat *S. aureus* infections. Preliminary results of bacteriophage isolation based on molecular epidemiology knowledge show that bacteriophages are specific of genetic lineages and that this strategy may be used as an approach to isolate and evaluate new bacteriophages for therapy.

Keywords: bacteriophage therapy, endolysins, enzybiotics, antibiotic resistance, molecular epidemiology

1. Introduction

1.1. *Staphylococcus aureus* as a zoonotic pathogen

Besides infecting human hosts in hospital-acquired (HA), in community-acquired (CA) infections as an opportunistic pathogen and in food poisoning by enterotoxic strains, *S. aureus* has also been isolated from animal hosts, both in livestock-associated (LA) and in companion animals' infections. Due to the raise of methicillin-resistant *S. aureus* (MRSA) strains, this feature was included as a phenotypic marker to identify *S. aureus*, and now they are described as MRSA or methicillin-sensitive *S. aureus* (MSSA). Molecular epidemiology approaches helped to the

Genetic lineage	Original described host	Further reports	Other features
ST1	Human	Cow, horse, chicken, pig[a]	–
CC5	Human	Chicken, turkey, dog[b,c]	ST5. Major HA clone; dog isolates in Japan and Spain
ST8	Human	Horse, cow, fish[d]	USA300. Major CA clone; fish isolates in Japan
ST9	Pig	Chicken	–
ST22	Human	Cat, dog[e,c]	EMRSA-15 global CA epidemic clone
CC97	Cow	Human, pig[f,g]	Loss and acquisition of virulence gene and pathogenicity islands lead to change in host specificity; recent transmission between cattle and pigs in Slovenia and Italy
ST121	Human	Rabbit	–
CC126	Cow	–	–
CC130	Cow	Sheep, deer[h]	In semiextensive red deer farm in Spain
CC133	Sheep	Goat, cow, cat[b], dog[c]	Cat isolates from Japan; dog isolates from Spain
ST239	Human	Cow	HA clone in Europe; isolates from bovine milk in Turkey[i]
CC705	Cow	–	–
CC385	Chicken	Wild birds	–
ST398	Pig	Human, cow, chicken, horse, dog[c]	Acquisition of genetic elements to evade immune response in new hosts. $mecA_{LGA251}$ (*mecC*); Spanish kennel dogs isolates
ST425	Cow	–	–
ST1464	Sheep	–	–

Modified from Refs. [5, 59, 60]; [a][61]; [b][62]; [c][63]; [d][64]; [e]ST deduced from homology between pet and human strains by PFGE and by *spa*-typing [65]; [f][7]; [g][8]; [h][66]; [i][67].

Table 1. Animal-associated genetic lineages of *S. aureus*.

understanding of the genetic structure of the *S. aureus* genetic population dynamics and hence in making predictions on transmissions between humans and animals. Multilocus sequence typing (MLST) is one of those molecular approaches. MLST analyzes the allelic combination of seven-to-nine (in *S. aureus* and other bacterial species) housekeeping genes that are randomly distributed along the genome. Mutations in *S. aureus* genes (*arcC, aroE, glpF, gmk, pta, tpi* and *yqiL*) are registered in an open public database (http://saureus.mlst.net) hosted at the Imperial College of London and supported by the Wellcome Trust Foundation. Each allele for each gene is designated with a specific number, so the allelic profile of a strain is designated by the numbers of alleles designated for each gene in the order described previously. Each allelic profile is designated with a sequence type (ST) number. STs sharing six or less alleles are grouped in clonal complexes (CC) in which the STs with the highest frequencies and number of shared alleles are designated as founder or subfounder clones, giving the name to the CC or related subgroups [1, 2]. Genetic lineages represented by a particular ST or CC are associated with specific hosts and geographical distributions. Some of them were originally described as specific for human or animal hosts and further reports associated them with animal or human transmissions, respectively, thus suggesting the zoonotic potential of *S. aureus* lineages. **Table 1** shows the major genetic lineages of *S. aureus* associated with animal hosts.

It is important to establish that the original description of a genetic lineage associated with a particular host followed by posterior reports of association with other hosts may not represent the evolutionary story of that lineage; it may only represent the original interest for the host due to the anthropocentric reasons or by the importance of the animal host as a food source or its contact with the human owner.

ST398 is one of the most reviewed cases of a clone showing animal-to-human transmission. Due to the whole-genome sequencing of strains from human endocarditis and bovine mastitis, differences in genomic content suggested that ST398 may be originated in humans. By loss, acquisition and reacquisition of pathogenicity islands or a staphylococcal chromosomal cassette related to methicillin resistance (SCC*mec*), and particular virulence genes like those encoding Panton-Valentine leukocidin (PVL) or the tetracycline resistance gene *tet*^{R}, ST398 susceptible to methicillin was originally transmitted from humans to animals and then back to humans as a methicillin-resistant strain [3, 4]. Similar events may occur for bovine-specific clones from CC97. Staphylococcal protein A (*spa*) and clumping factor A (*clfA*), which are important in human pathogenesis, appear as nonfunctional mutants in bovine isolates, suggesting that they are not important for bovine colonization. Alleles of von Willebrand factor are specific for each host, and pathogenicity islands seem also specific for each host [5,6].

Reports of interspecies transmission of *S. aureus* infections are becoming more frequent. In a study of CA-MRSA distribution in Slovenia, ST398, an originally pig-associated genotype, was found in 9.9% of the cases [7]. CC97 was first described as associated with bovine mastitis cases and now has also been found in humans and pigs. Of particular interest is the case of a multidrug-resistant LA-MRSA genotype from Italy that has been transmitted to pigs as MSSA and spilled back after methicillin resistance acquisition [8]. Ovine-associated *S. aureus* isolates are represented by CC133. In a global survey in Western Europe and Mediterranean countries, CC700 and CC522 were also ovine-associated. This distribution differs from North and South

America and Australia, where CC133 is the major ovine clone. Isolates from CC97 (bovine-associated), CC5, CC8 and CC30 (human-associated) were also found in this report, indicating high interspecific transmission of these genotypes [9]. Among zoological park animals in Greece, human-associated lineages ST80, ST8 and ST15, some of them with human pulsotype by PFGE analysis, suggest human-to-animal transmission [10]. ST80 and ST15 genetic lineages were also found in companion animals with close human contact in a veterinary teaching hospital in Greece. Panton-Valentine leukocidin (PVL), a necrotic toxin involved in skin infection, was found in 68.2% of MSSA isolates and in 50% of MRSA isolates, reinforcing the probable human origin of those strains. Also ST398 MRSA isolates were found that belong to the human cluster [11]. *S. aureus* has also been associated with wildlife animals. Studies in Spain demonstrate the presence of ST398 (pig- and human-associated) and ST1 (human-associated) MRSA isolates harboring the novel *mecC* methicillin resistance gene (see below) in either red deer, Iberian ibex, wild boar or Eurasian griffon vulture, suggesting a probable human origin of these isolates [12–14]. All of these examples represent the high transmission capability of apparently species-specific *S. aureus* genetic lineages and urge to the implementation of both molecular epidemiology surveillance and novel infection controls.

Antibiotic resistance is also a major problem of *S. aureus* infections. There is a constant interchange of mobile genetic elements modifying the virulence arsenal of *S. aureus* genetic lineages. This suggests that genetic background may be considered for the design of modern strategies to control *S. aureus* infections.

After the discovery of penicillin by Alexander Fleming in 1928 and its application to treat *S. aureus* infections in 1940, the first penicillin-resistant *S. aureus* strains were reported by 1945. Later in 1959, methicillin appears as an alternative to the use of penicillin. By 1961, the first methicillin-resistant *S. aureus* (MRSA) strains were reported. A similar story occurred for vancomycin-resistant (VRSA) and vancomycin-intermediate (VISA) *S. aureus*. Methicillin resistance is encoded by a staphylococcal chromosome cassette named SCC*mec* containing the *mecA* or *mecC* ($mecA_{LGA251}$) genes conferring resistance in humans and animals, for which at least 11 variants have been described. Apparently, these cassettes originated from a macrococcal *mecB* gene, which originated *mecA* (SCC*mec* and chromosomal forms) and *mecC* in staphylococci [15]. *mecC* has been almost exclusively associated with SCC*mec* type XI and located in animal strains from different STs and CCs [16], suggesting an intense intergeneric mobilization of SCC*mec* cassettes. VRSA strains seem not to be a major problem since only a dozen of clinical strains has been reported in the last decade. Vancomycin resistance is mediated by a complex of four genes (*vanA, vanH, vanX, vanY*) carried in a transposon. These modify a D-alanyl residue to D-lactate rendering the peptidoglycan structure resistant to vancomycin binding. *vanA* plasmids have also been reported, one of them being efficiently transmissible. This may predict that in the future, VRSA will also become a public health problem. Spontaneous mutants giving raise to VISA clones within vancomycin-susceptible *S. aureus* (VSSA) populations are known as heterogeneous-VISA (hVISA). hVISA/VISA is difficult to detect because on a first screening isolates behave as VSSA. Under the presence of vancomycin, VISA individuals are selected, and on a second screening, they behave as VISA. hVISA/VISA phenotypes have been associated with mutations in around 20 different genes that divert metabolism to peptidoglycan synthesis. Peptidoglycan then entraps vancomycin. hVISA/VISA reports are becoming more frequent in the literature, and it is to date considered of more relevance than VRSA. Staphylococci also present multidrug resistance genes such as *erm* (conferring resistance to macrolides, lincosamides and streptogramin B—MLS_B-) and *vga* (conferring resistance to lincosamides, pleuromutilins and streptogramin A) genes. Some of these genes are located in plasmids or transposons that are highly mobile genetic elements [16]. All of these evidences suggest that antibiotic resistance is becoming a major public health problem for the control of *S. aureus* infections, so alternative biotechnological approaches different from classical antibiotic treatments must be used in the future to control *S. aureus* infections. Bacteriophage therapy is one of those approaches.

1.2. Bacteriophages

Bacteriophages are viruses that infect only bacteria. They coevolve with their hosts optimizing its spread and release mechanisms from the bacterial cell to the environment and cause (in the case of lytic bacteriophages) lysis of the bacteria. They are also a major driving force in *S. aureus* evolution as a pathogen since many virulence genes are mobilized between different strains by means of transduction [17]. Bacteriophages are the most abundant biological entities of nature, although they are present in all environments, it is in aquatic systems where they are in greater proportion [18, 19]. Early indications of the presence of viral particles were reported in 1896 when bacteriologist Ernest Hanking observed that from the waters of the river Jumma in India, they identify a "substance" with antimicrobial activity against *Vibrio cholerae* and this substance was also heat labile and capable of passing through the filters of porcelain used at that time [20]. Two years later in 1898 Gamaleya observed a similar phenomenon in *Bacillus subtilis*. In 1915 and 1917, Twort and D'Herelle, respectively, discovered the viral particles called bacteriophages [21]. Frederick Twort in 1915 reported antimicrobial activity against *Staphylococcus aureus* suggesting that it could be viral particles among other possibilities. As of D'Herelle, he coined the term bacteriophage in 1917; this discovery was due to their previous studies to develop a vaccine against dysentery where he observed lytic plaques later named as bacteriophages [22]. In 1923, the National Institute of Bacteriophages in Tbilisi Georgia was established. Since then, the search for lytic bacteriophages for the biological control of infectious diseases has been in the scene.

1.3. Generalities

Bacterial viruses (bacteriophages or phages) possess genetic material in the form of DNA or RNA; morphologically, they consist of a head and a tail both constituted of protein. The head is the core package of nucleic acid surrounded by a protein shell or capsid also called lipo-protein. The tail varies on complexity from one bacteriophage type to another [23]. According to their lytic activity, they can be divided into two groups: lytic and lysogenic bacteriophages. When bacteriophages infect their host, they reproduce and the process ends with lysis of the bacteria and release of viral progeny. This is known as the lytic cycle. When the bacteriophages are able to integrate its genetic material into the bacterial genome and thus reproduce for several generations together with their host's genome, they are called temperate phages and they reproduce by a lysogenic cycle [24]. Bacteriophages which possess double-strand DNA express highly specific enzymes called viral-associated peptidoglycan hydrolyses (VAPGH) that bind to the bacterial cell surface and cause disruption of the cell wall to inject their DNA into the host cell [25]. The filamentous phage releases their viral progeny without causing the death of the bacteria [18], while nonfilamentous phages cause bacterial lysis by synthesizing endolysins (enzymes encoded by double-strand DNA phages) that hydrolyze peptidoglycan as part of an holin-endolysin system. The endolysins and holins are synthesized at late stages of phage infection. Endolysins accumulate in the cytoplasm until viral particles are assembled and holins form pores in the membrane allowing cytoplasmic translocation of endolysins through the membrane for peptidoglycan degradation [26]. Furthermore, single-stranded DNA or RNA bacteriophages synthesize "lysines" which interfere or inhibit the synthesis of the bacterial peptidoglycan [27]. The VAPGHs and endolysins are able to degrade the peptidoglycan when applied externally, which is why these enzymes represent an alternative to be used as enzybiotics in Gram-positive bacteria [28]. Bacteriophages and their endolysins are highly specific, infecting or hydrolyzing only a single species of bacteria attaching to specific receptors on the surface of host cell. The specificity of interaction between phage attachment structures and host cell surface receptors determinates host range. [29].

2. Bacteriophage reproduction

2.1. Lytic cycle

Phages replicate inside bacterial host and the process finalizes with lysis of the host and spreading of phage progeny. Phage replication includes the following steps [30]:

1. **Adsorption**. Phage attachment to a specific host cell in a process involving interaction with receptors on the surface of a susceptible host cell and an infecting virus. There are two major types of receptors: components of a bacterial cell like lipopolysaccharide, peptidoglycan, outer membrane proteins and teichoic acids, and fimbriae-type receptors like *pilli* or flagella.

2. **Nucleic acid injection**. Through the tail, phage injects its genetic material into the cell after peptidoglycan degradation behind pore formation (by VAPGH). The phage coat protein that includes capping head and tail structure remains attached to the bacterial surface.

3. **Replication**. After injection of its nucleic acid, phage expresses early genes that redirect host synthesis machinery to the reproduction of viral nucleic acid and proteins.

4. **Assembly and packing phage particles**. Once the viral components are synthesized, the genetic material is encapsulated in its protein coat, and complete virus particles are formed.

5. **Phage progeny release**. Phage late proteins like holins and endolysins or murein synthesis inhibitors are produced, and they are responsible for the lysis of the host cell and the release of viral particles to the environment.

http://dx.doi.org/10.5772/65761

2.2. Lysogenic cycle

The lysogenic cycle comprises the same steps as lytic cycle, but after penetration of the genetic material, the phage nucleic acid is inserted into the chromosome of the bacteria and is replicated as a segment of the own bacterial genome for one or more generations without metabolic consequences for the bacterium. After this cycle, the genetic material of the phage can be excised from the bacterial chromosome and enter into a lytic cycle; usually, this occurs under physiological stress or damage of the genetic material.

3. Endolysins

The term endolysin was coined until 1958 to refer to the phage component responsible for the bacterial lysis. Lytic phages present a genetic cassette encoding a holin-endolysin system. At the end of the reproductive cycle, once mature viral particles have been assembled, holins are synthesized in critical concentrations and inserted into the cell membrane, creating pores for the translocation of endolysins, previously accumulated in the cytoplasm, to reach the peptidoglycan structure [19]. Endolysins are classified according to its enzymatic activity (**Figure 1**) in: (1) N-acetylmuramoyl-alanine amidases, which hydrolyze the amide bond

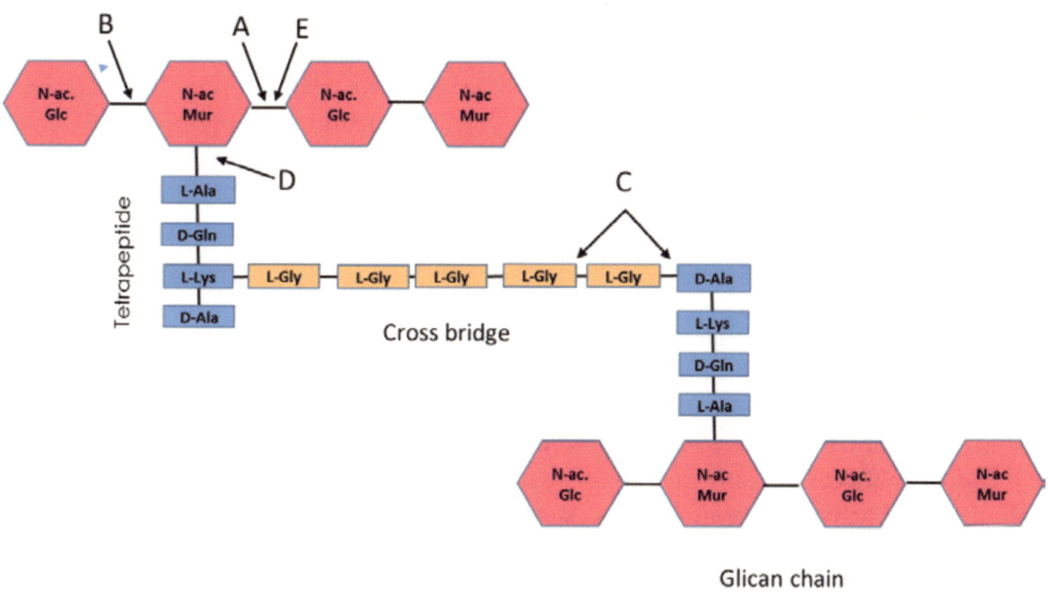

Figure 1. Enzymatic activities of endolysins. (A) N-acetyl-muramidase catalyzes the hydrolysis of N-acetylmuramoil-β-1,4-N-acetylglucosamine. (B) N-acetylglucosaminidase catalyzes the hydrolysis of N-acetylglucosaminil-β-1,4-N-acetylmuramine. (C) Endopeptidase hydrolyzes peptidic bonds on amino acids chains linked to the glycan moiety or in the pentapeptidic bridge. (D) N-acetylmuramoyl-L-alaninamidase hydrolyzes the amide bond that connects the glycan with the amino acids. (E) Transglycosylases attach the glycosidic β-1,4 bonds resulting in the formation of a 1,6 anhydrous ring in N-acetylmuramic acid (modified from Barrera-Rivas et al. [19]).

between the N-acetyl-muramic in the glycan chain and the L-alanil residues; (2) endo-β-N-acetylglucosaminidases, which hydrolyzes the N-acetylglucosamine β-1,4-N-acetylmuramine acid linkage; (3) N-acetyl-muramidases, which catalyze the hydrolysis of N-acetylmuramoil-β-1,4-N-acetilglucosamine bond; (4) transglycosylases, which disrupt β-1-4 glycosidic bonds by forming a 1–6 anhydride ring in the N-acetylmuramic residue; (5) endopeptidases, which may hydrolyze both the tetrapeptide linked to the glycosil moieties and the pentapeptide entrecrossing bridge [31, 32].

Endolysins encoded by double-stranded DNA bacteriophages have a molecular weight between 25 and 40 kDa [33]. Most of endolysins are composed of at least two functional domains: one containing the catalytic activity located generally in the N-terminal domain and one responsible for the recognition of a specific substrate associated with the C-terminal domain. In some cases, more than one catalytic domain or more than one recognition domain are present [19]. The recognition domain usually joins to specific molecules in the bacterial cell envelopes such as monosaccharides, coline or teichoic acids [34]. Endolysin activity is usually species specific, although there have been reports of endolysins with a wider substrate range. Besides, the cell wall recognition domain is not always essential for endolysin activity. The endolysin got a wider substrate range, but it conserved certain specificity, since it was no active against all bacteria. Studies of crystallography and mutation analysis with endolysin *PlyL* against *Bacillus anthracis* led to propose that the C-terminal domain of this endolysins inhibits the activity of the catalytic domain by particular intermolecular interactions. This inhibition is released when the C-terminal domain binds to its particular ligands in the target cell wall, thus acting as a regulatory domain [35]. Most of the reported endolysins from phages against *S. aureus* have two catalytic domains and a cell wall recognition domain being LysK one of the must studied endolysin models. LysK has a cysteine/histidine-dependent aminohydrolase/peptidase (CHAP) catalytic domain that hydrolyzes the peptidic bond between the D-alanine of the oligopeptide chain attached to the sugar backbone and the first glycine of the pentaglycine bridge that is typical of *S. aureus* peptidoglycan and confers resistance to lysozyme. CHAP presents the higher activity of both hydrolytic domains. LysK also has an N-acetylmuramoyl L-alanine amidase or amidase-2 (Ami-2) catalytic domain which catalyzes the hydrolysis of the N-glycosidic bond between the N-acetylmuramic residue and the L-alanine of the oligopeptide attached to the sugar backbone. A third domain called SH3b is responsible for the specific recognition of cell wall components, strain specificity and modulator of hydrolytic activities [36, 37]. Endolysin 2638A has similar triple domain structures: an amino-terminal domain with endopeptidase activity, a central Ami-2 domain (with the highest activity in this phage) and a SH3b cell wall recognition domain [38]. Modular structure of *S. aureus* endolysins has allowed the construction of chimeric endolysins by the combination of catalytic and/or recognition domains. An example is the endolysin Ply187AN-KSH3b, which is a translational fusion of the CHAP domain of phage Ply187 and the cell wall recognition domain SH3b from LysK endolysin. This endolysin was effective in a mouse model of endophthalmitis that also decreased inflammatory response and protected the retina from tisular damage [39].

http://dx.doi.org/10.5772/65761

4. Evolution of phage therapy

Since the discovery of bacteriophages, it raised the idea of using them for treatment of bacterial infections. D'Herelle began testing the therapeutic effects of phages, using animal models such as chickens and cows first, which provided successful results. Subsequently, there was carried out human testing and the development of phage therapies became more extensive. In 1923, the development of phage-based therapy strengthened with the foundation of the Eliava Institute in Tbilisi, Georgia, in the former USSR. In 1940, they began to commercialize phage in the United States. During World War II, phage cocktails were used to treat diseases such as dysentery and gangrene in the soldiers of the former Soviet Union. Their application was topical, oral and intravenous, although the latter favors the immune response of the individual treated due to the protein content of the virus, resulting in the elimination of the phage from the body [40–42]. Until a few years ago, therapies were based solely on the administration of the complete bacteriophage, but it was until 2000 that the studies for the identification and purification of lytic enzymes to treat infections caused by bacteria begun. In addition to using bacteriophages and their enzymes as enzybiotics (enzymatic activities with antibiotic effect) in the treatment of infections in humans, animals and agriculture, they are also used in the food industry as preservatives and disinfectants [19]. After the discovery of penicillin, the development and commercialization of antibiotics in the 1940s and 1950s soon occupied the global antibacterial market. The lack of knowledge of the biology of phages, the lack of studies of epidemiology of diseases and also a lack of control during the preparation of therapeutic

Bacteriophages	**Antibiotics**
Advantages • More abundant entities in nature • They are natural enemies of bacteria • Ecologically friendly • Don't affect normal microbiota • Bacteria don't develop resistance	• Doses are easy to determine • Broad spectrum of action for the treatment of several infections; immediately used without identifying the specific strain causing the infection.
Disadvantages • Just a small number of phages are effective as therapeutic agents • It is necessary to identify the specific strain causing the infection to use the specific and active phage	• Production of synthetic or semisynthetic antibiotics can contaminate environment • Destroy all bacteria cells including normal microbiota • Bacteria develop resistance

Table 2. Advantages of bacteriophages over antibiotics.

stocks led to a temporary delay in the research and development of phage therapy. In early studies of phage preparations, successful results showed high antimicrobial activity in *in vitro* and *in vivo* assays; however, in subsequent trials, some phages had little or no ability to destroy bacteria or became lysogenic [43].

Because of the concern in the treatment of diseases caused by pathogens with multiple resistance to antibiotics, it has revived the interest in the development and use of the bacteriophage therapy and their enzymes to treat diseases in animals and humans. Phage therapy has been used in plants, animals and humans with varying degrees of effectiveness; in addition, bacteriophages have some potential advantages over antibiotics but also have some disadvantages [44] (**Table 2**). The specificity of phage-host interaction permits the use of some phages in therapy because they do not have influence on normal microbiota in humans, animals, plants, food or inert surfaces. On the contrary, the use of broad spectrum antimicrobials has an effect on the eradication of a wide range of infecting pathogens but also kills bacteria from the natural microbiota thus causing a disequilibrium in the host normal microbiota and promotes secondary bacterial or fungi infections or even physiological or endocrinological disorders.

5. Bacteriophages and its interaction with animals

There is a high diversity of phages in microbial communities living in symbiosis with animals, for example, in the pig digestive tract and in the cow rumen [45, 46]. In the animals gut microbiota, there is a complex ecosystem with approximately 500 species of microorganisms, which are interacting with mutual benefits [47]. When the abundance of one of those bacteria changes and alters the dynamic equilibrium, it results in some disorders or disease in the host. Phages play an important ecological role for the health regulating the relative amount of the different bacterial strains in microbiota. On the other hand, the presence of phages in animals could present some disadvantages for health. When phages insert into the bacterial genome genes that encode toxins like Panton-Valentine, Shiga and diphtheria toxins [48, 49] or some other virulence factors, further excision may be aberrant, leading the phage genome to carry those virulence genes by transduction. These aberrant phages may insert in new hosts and transfer virulence properties. In fact, some genetic elements related to virulence may be originated from aberrant prophages. Also, prophages confer its host resistance to the infection of other phages. In addition, phages can also impact in host immune response through modifications in bacteria's antigenicity. Density of host bacteria determines the ability of phages to infect and reproduce because phages encounter their host through random collision. There are four models in the literature explaining the behavior of phages and bacteria in the regulation of animal microbiota. (A) "**kill the winner**": phages are more abundant than bacteria but don't infect them because of the lower abundance of its host, when some strains overgrow, phages can depredate and kill them by lysis, and system comes back to an initial healthy equilibrium. (B) "**kill the relative**": some phages are reproduced from lysogenic strains so they don't need to be abundant; strains with prophages produce phages that kill their genetically related strains which aren't resistant to the phage. The result is an advantage in the abundance of lysogenic population in comparison with non-lysogenic strains. (C) "**community shuffling**": temperate phages act negatively on their host, temperate phages kill their host under some stress situation and this don't occur with non-lysogenic strains. Positive feedback could take place if massive lysis causes host reactions like inflammation on another immune response. This causes an imbalance in the microbiota and in some cases disorders or diseases related to the change in populations. (D) "**invade the relative**": prophage propagates itself by infecting new hosts without lysing them, but establishing lysogeny [47]. Other contribution of phages to bacteria strains in animal microbiota is when phages function as vectors of virulence, for example, changing the expression of antigens in outer membrane like O-antigens [50], giving to bacteria genetic adaptation; it results in new and more virulent strains for the animal host.

6. Phage therapy in animal infections caused by *S. aureus*

The use of bacteriophages or bacteriophage cocktails and the use of endolysins represent a potential alternative for the treatment of infections caused by *S. aureus*. Although several diseases caused by *S. aureus* in animals have been described in a previous section, most of the research in phage therapy has been done for bovine mastitis, so it will be the central point of this section.

7. Mastitis and *S. aureus*

Mastitis is characterized by the inflammation of the mammary gland in one or more quarters of the udder accompanied of leukocyte production, mainly monocytes and blood serum proteins such as cytokines, chemokines and interleukins [51]. It is caused mostly by contagious pathogens such as *S. aureus* and *Streptococcus* spp. and environmental pathogens such as *E. coli*. Also, in less proportion, mastitis can be caused by or promoted by injury, allergies and neoplasias [52]. Mastitis causes large economic losses in the milk and dairy products industry for about 2 billion of dollars each year in the USA [53]. Among the pathogens causing mastitis, *Staphylococcus aureus* is considered a causal agent of great concern because of the low cure rate of *S. aureus* infections by antibiotic treatment and its ability to persist in a herd in the form of undetected subclinical infections [54]. Vaccines for the treatment of mastitis have limited efficacy. Cure rates for antibiotic treatment are often lower than 15%. This is caused by the poor penetration of the gland by antibiotics allowing *S. aureus* to survive inside the epithelial or phagocytic cells. Antibiotic resistance in *S. aureus* is also a growing concern, with overall rates of antimicrobial resistance in bovine *S. aureus* isolates varying widely by region [55]. The continued emergence of MRSA strains in humans and animals points to the need to develop new antimicrobial agents or therapies treatment for this pathogen. The treatment of bacterial infections with bacteriophages and their derivatives is such an option. **Table 3** describes those approaches.

Experiment	Observations/treatment	Reference
Bacteriophages		
Use of phage K to treat cow with subclinical mastitis. Twenty-four lactating Holstein cows with pre-existing subclinical *S. aureus* mastitis were treated. Prior to experimentation with dairy cows, the phage preparations were screened in mice to determine acute toxic effects	Treatment consisted of 10 ml intramammary infusions of 1.25×10^{11} PFU of phage K and infusions with saline for control, administered once per day for 5 days. The cure rate was established by the assessment of four serial samples collected following treatment The cure rate was 3 of 18 quarters (16.7%) in the phage-treated group, whereas none of the 20 saline-treated quarters were cured which were already infected with *S. aureus*. Phage-infused healthy quarters continued to shed viable bacteriophage into the milk for up to 36 h postinfusion	[54]
Study of bacteriophage (M^{Sa}) active against *Staphylococcus aureus*, including methicillin-resistant staphylococcal strains	A lethal dose of *S. aureus* A170 was given to mice; phage M^{Sa} rescued 97% of mice and completely eradicated bacteria in vivo within 4 days of phage treatment; when applied to nonlethal (5×10^6 CFU/mouse) 10-day infection, the phage also fully cleared the bacteria The phage M^{Sa}, delivered inside macrophages by *S. aureus*, kills the intracellular staphylococci *in vivo* and *in vitro* Phage M^{Sa} was well tolerated by the animals, it drastically reduced inflammation, and it did not stimulate the production of neutralizing antibodies	[68]
Isolation of a novel virulent bacteriophage (MSA6) from a cow with mastitis	Isolated phage was capable of infecting a wide spectrum of staphylococcal strains of both human and bovine origin	[69]
Isolation of bacteriophages virulent against *Staphylococcus aureus* associated with goat mastitis. Bacteriophages were isolated from soil and fecal samples	Three of the bacteriophage isolates, phage/CIRG/1, phage/CIRG/4 and phage/CIRG/5, exhibited lytic activity against over 80% of the staphylococcal isolates. All isolates were stable up to 3 months at 37°C, and for 16 months at 4°C but the stability of their respective endolysins only lasted for 12–23 days at 37°C and 6 months at 4°C. Lytic activity was determined *in vitro*	[70]
Isolation of a phage that infects *S. aureus* from bovine mastitis. SA phage was isolated from sewage water	Authors analyzed *in vitro* the sensibility to phage infection of five *S. aureus* strains with drug resistance. Phages were stable at wide temperature and pH ranges SA phage efficiently reduced bacterial growth in the bacterial reduction assay	[71]
Endolysins		
Fusion of endopeptidase domain from streptococcal endolysin SA2 with either lysostaphin or LysK endolysin and the recognition domain of endolysin LysK	In a mouse model of mastitis, chimeric SA2-E-Lyso-SH3B and SA2-E-LysK-SH3B reduce *S. aureus* CFUs by 1–3 log units in cow milk and by 0.63–0.8 log units in mammary glands. Synergism with lysostaphin reduced CFUs by 3.36 log units	[72]

Experiment	Observations/treatment	Reference
Engineering triple-acting staphylolytic peptidoglycan hydrolases. Both amidohydrolase/peptidase and amidase domains from LysK bacteriophage fused with the N-terminal domain of lysostaphin	Modification of the triple-acting lytic construct with a protein transduction domain significantly enhanced both biofilm eradication and the ability to kill intracellular *S. aureus* as demonstrated in cultured mammary epithelial cells and in a mouse model of staphylococcal mastitis shows that bacterial cell wall degrading antimicrobial enzymes can be engineered to enhance their value as potent therapeutics	[73]
Endolysin gene from novel bacteriophage IME-SA1 expressed in pET-32a fused with Trx-SA1	Each udder quarter suffering from mild clinical mastitis received the experimental treatment of intramammary infusion of 20 mg of recombinant endolysin once per day. Milk samples were taken on days 1, 2 and 3 from each infected udder quarter before treatment for SCC determination and microbiological analysis. Preliminary results of therapeutic trials in cow udders showed that Trx-SA1 could effectively control mild clinical mastitis caused by *S. aureus*	[74]

Table 3. Bacteriophages and endolysins therapy for treatment of *S. aureus* mastitis.

8. Animal models for treatment of other *S. aureus* infections

Animal models have been widely used to evaluate the performance of phage therapy in the treatment of a variety of infections caused by *S. aureus*, usually nosocomial infections in humans. **Table 4** presents the use of phages and/or their endolysins in infections by *S. aureus* in animal models.

Experiment	Observations/treatment	Reference
Bacteriophages		
Isolation of φMR11 phage, tested against *S. aureus* in mice causing bacteremia	Intraperitoneal administration of purified φMR11 can protect mice with bacteremia caused by methicillin-resistant *S. aureus*. Use of φMR11 did not cause any adverse effects	[75]
Isolation of Stau2 phage from hospital effluents. Tested in mice infected with *S. aureus* S23	*S. aureus* inoculated in an injection with 0.5 ml in intraperitoneal cavities of the mice. Protection by Stau2 from a lethal bacterial infection occurred in a dose-dependent manner. Immediate phage administration provided better protection than delayed administration. The surviving mice remained healthy during the 14-day observation period. Injection with a large amount of phage (7.5×10^{10} PFU) or SM buffer alone did not affect their physical condition during the same period. Injection with a mechanical bacterial lysate of strain S23 did not protect the mice from a lethal infection,	[76]

Experiment	Observations/treatment	Reference
	suggesting that bacterial components, such as bacteriocins, were not involved in protection	
S13 phage against lung-derived lethal septicemia by *S. aureus* strain SA27 in a mouse model.	Intranasal application of *S. aureus* strain SA27 induced 93% lethality in 3 days. S13 phage was done administered 6 h postinfection with 0.2 ml of solution of 15 × 10^{10} PFU/ml. The survival rates of phage administered and control groups were 67% and 10% on day 5, respectively. The administration of phage S13 reduced the *S. aureus* cell densities with significant phage replication in different tissues and it rescued the infected mice	[77]
Endolysins		
Endolysin LysGH15 derived from staphylococcal phage GH15 was used against MRSA *in vivo* using mice and *in vitro*	Mice were infected with 2× of the minimum lethal dose of MRSA. The bacterial growth in spleens was determined 1–24 h after the lethal infection. Although the number of bacteria in spleens decreased slightly 6–12 h after infection, it increased until death. In contrast, the number of MRSA cells in spleens declined by 2 log units at 5 h after LysGH15 treatment (50 μg/mouse) in the lethal MRSA-infected mice and continue decreasing to reach an undetectable level. Also, LysGH15 treatment could modulate inflammation reducing the levels of IL-6, IL-4 and IFN-γ mRNA in spleens	[78]
PlySs2 bacteriophage lysine derived from *Streptococcus suis* was used to treat MRSA which cause bacteremia in mice	Mice were infected i.p. with MRSA (MW2). PlySs2 protected mice and result in 89% survival in a bacteremia model, while in the control group without treatment with PlySs2 only 6% of mice survived	[79]
Nine endolysins within an homology group sharing SH3b domain but diverse classes of peptidoglycan hydrolyses (PGHs) from *S. aureus* were tested to determinate their antimicrobial activity	Proteins were expressed, purified and tested for staphylococcal activity *in vitro*. Cut sites from endolysins were determined. PGHs show different degrees of activity *in vitro*. Some PGHs can eliminate biofilms. Six of the nine PGHs protected from death at 100% of infected mice with MRSA	[80]

Table 4. Use of phages and endolysins against *S. aureus* infections using animals models.

9. A functional molecular epidemiology approach to isolate bacteriophages against specific genetic lineages of *S. aureus*

As stated previously, particular genetic lineages are related to host specificity and pathogenic strategies of *S. aureus*. In a previous work, we isolated and typed *S. aureus* isolates from bovine mastitis in backyard farms in México. Most of these isolates were related to CC5 subgroups ST97 and ST126 and present diverse *spa*-types [56,57]. An isolate of ST8 (CA, human-associated) genetic background was also found. Several isolates from different STs were selected according to their *spa*-type or their antimicrobial resistance profile. **Table 5** shows examples of phages isolated using the selected molecular-typed *S. aureus* strains. All of the strains used for isolation belong to CC5 subgroup 97, but differed in their *spa*-type and their resistance profile, or belong to the ST8. Twenty-eight bacteriophages were isolated from 10 different *S. aureus* genetic lineages. Host ranges of isolated phages included strains from the same genetic lineage (CC5 subgroup 97). NST-1 corresponds to a new ST that is a single locus variant of ST126. None of these bacteriophages were active against the isolate with ST8 genotype. Restriction fragment length polymorphism with *XbaI* enzyme revealed only four different phage genotypes (data not shown). Phages MICHSAF5 and MICHSAF9 were clustered in the same RFLP group, whereas MICHSAF1 and MICHSAF15 were from different groups.

Phage	Strain for isolation	Genotype (ST and *spa*-type) and antibiotic resistance	Susceptible STs	Susceptible CCs
MICHSAF1	MRI-166	ST352/t267/GM	97, 352	5 subgroup ST97
MICHSAF5	MRI-150	ST97/t4570/NB, GM, FOX	126, NST-1, 97, 352	5 subgroups ST97 and ST126
MICHSAF9	MRI-150	ST97/t4570/NB, GM, FOX	126,NST-1, 97, 352	5 subgroups ST97 and ST126
MICHSAF15	MRI-151	ND/ND/NB, GM, FOX, C, CC, L, E, LZD	97, 352	5 subgroup ST97

NB, novobiocin; GM, gentamicin; FOX, cefoxitin; C, chloramphenicol; CC, clindamycin; L, lincomycin; E, erythromycin; LZD, linezolid; ND, not determined.

Table 5. Preliminary analysis of phage isolation using a molecular typing background.

It is interesting to note that phages MICHSAF5 and MICHSAF9 were isolated using the same strain as host, and both presented the same host range and RLFP pattern. Phages MICHSAF1 and MICHSAF15 were associated with strains with different STs and resistance patterns, and the genotypes of the susceptible *S. aureus* strains were similar. All strains used for the isolation of bacteriophages and the susceptible strains belonged to CC5 subgroups ST97 and ST126. These results suggest that genetic background of the strain used for isolation of the bacteriophage will determine the host range of the bacteriophage.

10. Conclusions

Bacteriophages and their endolysins in its natural or recombinant forms have proven to function in animals and animal models to control diverse forms of *S. aureus* infections. More structure-function studies of endolysins will contribute to design recombinant enzybiotics for the control of *S. aureus* infections. Functional molecular epidemiology is the applied use of the knowledge generated by molecular epidemiology to establish strategies for the control of infectious diseases [58] such as bacteriophage therapy. Bacteriophage selection using finely typed strains will help to properly select phages for therapy and to analyze the host range of the isolated bacteriophages. The strains typed by molecular approaches may also be useful to test ranges of activity of phage-derived endolysins. These, along with genetic engineering for the study and expression of endolysins, will help to design better biotechnological approaches for the control of infectious diseases.

Acknowledgements

This work was supported by Consejo Nacional de Ciencia y Tecnología, México, Fondos Mixtos-Michoacán MICH-2012-C05-196879 and Coordinación de la Investigación Científica de la Universidad Michoacana de San Nicolás de Hidalgo, México, Grant 14.7. C.I. Barrera-Rivas and N.A. Valle-Hurtado received scholarships for MSc. studies from Consejo Nacional de Ciencia y Tecnología, México.

Author details

Claudia I. Barrera-Rivas[1], Norma A. Valle-Hurtado[1], Graciela M. González-Lugo[2], Víctor M. Baizabal-Aguirre[1], Alejandro Bravo-Patiño[1], Marcos Cajero-Juárez[3] and Juan J. Valdez-Alarcón[1*]

*Address all correspondence to: jjvaldez_a@hotmail.com and jjvaldez@umich.mx

1 Multidisciplinary Center of Biotechnology Studies, Faculty of Veterinary Medicine and Animal Production, Michoacana University of San Nicolás of Hidalgo, Tarímbaro, Michoacán, México

2 Department of Microbiology, Biological Sciences National School, Polytechnical National Institute, Ciudad de México, México

3 Multidisciplinary Center of Biotechnology Studies, Faculty of Veterinary Medicine and Animal Production, and Institute of Agriculture, Livestock and Forestry Research, Michoacana University of San Nicolás of Hidalgo, Tarímbaro, Michoacán, México

References

[1] Enright MC, Day NPJ, Davies CE, Peacock SJ, Spratt BG: Multilocus sequence typing for characterization of methicillin resistance and methicillin susceptible clones of *Staphylococcus aureus*. J. Clin. Microbiol. 2000;38:1008–1015.

[2] Maiden MCJ: Multilocus sequence typing of bacteria. Ann. Rev. Microbiol. 2006;60:561–568.

[3] Fluit AC: Livestock-associated *Staphylcoccus aureus*. Clin. Microbiol. Infect. 2012;18:735–744. doi:10.1111/j.1469-0691.2012.03846.x

[4] Price LB, Stegger M, Hasman H, Aziz M, Larsen J, Andersen PS, Pearson T, Waters AE, Foster JT, Schupp J, Gillece J, Driebe E, Liu CM, Springer B, Zdovc I, Battisti A, Franco A, Zmudzki J, Schwarz S, Butaye P, Jouy E, Pomba C, Porrero MC, Ruimy R, Smith TC, Robinson DA, Weese JS, Arriola CS, Yu F, Laurent F, Keim P, Skov R, Aarestrup FM: *Staphylococcus aureus* CC398: host adaptation and emergence of methicillin resistance in livestock. MBio. 2012;3:e00305–11. doi:10.1128/mBio.00305-11

[5] Fitzgerald JR: Livestock-associated *Staphylococcus aureus*: origin, evolution and public health threat. Trends Microbiol. 2012;20:192–198. doi:10.1016/j.tim.2012.01.006

[6] Lindsay JA: Genomic variation and evolution of *Staphylococcus aureus*. Int. J. Med. Microbiol. 2010;300:98–103. doi:10.1016/j.ijmm.2009.08.013

[7] Dermota U, Mueller-Premru L, Svent-Kucina N, Petrovic Z, Ribic H, Rupnik M, Janezic S, Zdovk I, Grmek-Koznik I: Survey of community-associated methicillin-resistant *Staphylococcus aureus* in Slovenia: identification of community-associated and livestock-associated clones. Int. J. Med. Microbiol. 2015;305:505–510. doi:10.1016/j.ijmm.2015.05.002

[8] Feltrin F, Alba P, Kraushaar B, Ianzano A, Argudín MA, Di Matteo P, Porrero MC, Aarestrup FM, Butaye P, Franco A, Battisti A: Molecular characterization of a livestock-associated, multidrug-resistant, methicillin-resistant *Staphylococcus aureus* clonal complex 97 lineage spreading in dairy cattle and pigs in Italy. Appl. Environ. Microbiol. 2015;82:816–821. doi:10.1128/AEM.02854-15

[9] Smith EM, Needs PF, Manley G, Green LE: Global distribution and diversity of ovine-associated *Staphylococcus aureus*. Infect. Genet. Evol. 2014;22:208–215. doi: 10.1016/j.meegid.2013.09.008

[10] Drougka E, Foka A, Posantzis D, Giormezis N, Anasstassiou ED, Petinaki E, Spiliopoulou I: Human *Staphylococcus aureus* lineages among zoological park residents in Greece. Open Vet. J. 2015;5:148–153. PMID:26623381

[11] Drougka E, Foka A, Koutinas CK, Jelastopolu E, Giormezis N, Farmaki O, Sarrou S, Anasstassiou ED, Petinaki E, Spiliopoulou I: Interspecies spread of *Staphylococcus aureus* clones among companion animals and human close contacts in a veterinary teaching hospital. A cross-sectional study in Greece. Prev. Vet. Med. 2016;126:190–198. doi:10.1016/j.prevetmed.2016.02.004

[12] Porrero MC, Mentaberre G, Sánchez S, Fernández-LLario P, Gómez-Barrero S, Navarro-González N, Serrano E, Casas-Díaz E, Marco I, Fernández-Garayzabal JF, Mateos A, Vidal D, Lavín S, Domínguez L: Methicillin-resistant *Staphylococcus aureus* (MRSA) carriage in different free-living wild animal species. Vet. J. 2013;198:127–130.

[13] Porrero MC, Valverde A, Fernández-Llario P, Diez-Guerrier A, Mateos A, Lavin S, Cantón R, Fernández-Garayzabal JF, Domínguez L: *Staphylococcus aureus* carrying *mecC*

gene in animals and urban wastewater, Spain. Emerg. Infect. Dis. 2014;20:899–901. doi: 10.3201/eid2005.130426

[14] Porrero MC, Mentaberre G, Sánchez S, Fernández-Llario P, Casas-Díaz E, Mateos A, Vidal D, Lavín S, Fernández-Gayarzabal JF, Domínguez L: Carriage of *Staphylococcus aureus* by free-living wild animals in Spain. Appl. Environ. Microbiol. 2014;80:4865–4870. doi:10.1128/AEM.00647-14

[15] Hiramatsu K, Katayama Y, Matsuo M, Sasaki T, Morimoto Y, Sekiguchi A, Baba T: Multi-drug-resistant *Staphylcoccus aureus* and future chemotherapy. J. Infect. Chemother. 2014;20:593–601. doi:10.1016/j.jiac.2014.08.001

[16] Becker K, Ballhausen B, Köck R, Kriegeskorte A: Methicillin resistance in Staphylococcus isolates: the "mec alphabet" with specific consideration of mecC, a mec homolog associated with zoonotic *S. aureus* lineages. Int. J. Med. Microbiol. 2014;304:794–804. doi:10.1016/j.ijmm.2014.06.007

[17] Xia G, Wolz C: Phages of *Staphylococcus aureus* and their impacto on host evolution. Infect. Genet. Evol. 2014;21:593–601. doi:10.1016/j.meegid.2013.04.022

[18] Bergh O, Børsheim KY, Bratbak G, Heldal M: High abundance of viruses found in aquatic environments. Nature 1989;340:467–468.

[19] Barrera-Rivas CI, Cajero-Juárez M, Oviedo-Boyso J, Nuñez-Anita RE, Kameyama-Kawabe L, Valdez-Alarcón JJ: Advances in the use of endolysins: general remarks, structure, applications, genetic modifications and perspectives. In: Mendez-Vilas A, Editor. The battle against microbial pathogens: basic science, technological advances and educational programs. Vol. 1. Spain: Formatex Research Center. p. 259–268. ISBN: 978-84-942134-5-8

[20] Sulakvelidze A, Alavidze Z, Morris JG: Bacteriophage therapy. Antimicrob. Agents Chemother. 2001;45:649–659.

[21] Duckworth DH: Who discovered bacteriophage? Bacteriol. Rev. 1976;40:793–802.

[22] Summers WC: Bacteriophage discovered. Felix D'Herelle and the Origins of Molecular Biology. Yale University Press. 1999;1:47–59.

[23] Wang CH, Bau HJ, Yeh SD: Comparison of the nuclear inclusion b protein and coat protein genes of five papaya ringspot virus strains distinct in geographic origin and pathogenicity. Phytopathology, 1994, 84:1205–1209.

[24] Inal JM: Phage therapy: a reappraisal of bacteriophages as antibiotics. Arch. Immunol. Ther. Exp (Warsz). 2003;51:237–244.

[25] Rodríguez-Rubio L, Martínez B, Donovan DM, Rodríguez A, García P: Bacteriophage virion-associated peptidoglycan hydrolases: potential new enzybiotics. Crit. Rev. Microbiol. 2013;39:427–434.

[26] Wang IN, Smith DL, Young R: Holins: the protein clocks of bacteriophage infections. Ann. Rev. Microbiol. 2000;54:799–825. doi:10.1146/annurev.micro.54.1.799

[27] Borysowski J, Weber-Dabrowska B, Górski A: Bacteriophage endolysins as a novel class of antibacterial agents. Exp. Biol. Med. 2006;231:366–377.

[28] Nelson D, Loomis L, Fischetti VA: Prevention and elimination of upper respiratory colonization of mice by group A streptococci by using a bacteriophage lytic enzyme. Proc. Natl. Acad. Sci USA. 2001;98:4107–4112. doi:10.1073/pnas.061038398

[29] Weinbauer MG, Rassoulzadegan F: Are viruses driving microbial diversification and diversity? Environ. Microbiol. 2004;6:1–11.

[30] Drulis-Kawa Z, Majkowska-Skrobek G, Maciejewska B, Delattre AS, Lavigne R: Learning from bacteriophages-advantages and limitations of phage and phage-encoded protein applications. Curr. Protein Pept. Sci. 2012;13:699–722.

[31] Schmelcher M, Donovan DM, Loessner MJ: Bacteriophage endolysins as novel antimicrobials. Future Microbiol., 2012;7:1147–1171. doi:10.2217/fmb.12.97

[32] Loessner MJ: Bacteriophage endolysins—current state of research and applications. Curr. Opin. Microbiol. 2005;8:480–487. doi:10.1016/j.mib.2005.06.002

[33] Fischetti VA: Bacteriophage endolysins: a novel anti-infective to control Gram-positive pathogens. Int. J. Med. Microbiol. 2010;300:357–362. doi:10.1016/j.ijmm.2010.04.002

[34] Sass P, Bierbaum G: Lytic activity of recombinant bacteriophage ϕ11 and ϕ12 endolysins on whole cells and biofilms of *Staphylococcus aureus*. Appl. Environ. Microbiol. 2007;73:347–352. doi:10.1128/AEM.01616-06

[35] Low LY, Yang C, Perego M, Osterman A, Liddington RC: Structure and lytic activity of a *Bacillus anthracis* prophage endolysin. J. Biol. Chem. 2005;280:35433–35439. doi:10.1074/jbc.M502723200

[36] Horgan M, O'Flynn G, Garry J, Cooney J, Coffey A, Fitzgerald GF, Ross RP, McAuliffe O: Phage lysine LysK can be truncated to its CHAP domain and retain lytic activity against live antibiotic-resistance *Staphylococcus aureus*. Appl. Environ. Microbiol. 2009;75:872–874. doi:10.1128/AEM.01831-08

[37] Sanz-Gaitero M, Keary R, García-Doval C, Coffey A, van Raiij MJ: Crystallization of the CHAP domain of the endolysin from *Staphylococcus aureus* bacteriophage K. Acta Crystallogr. Sect. F Struct. Biol. Cryst. Commun. 2013;69:13913–1396. doi:10.1107/S1744309113030133

[38] Abaev I, Foster-Frey J, Korobova O, Shishkova N, Kiseleva N, Kopylov P, Pryamchuk S, Schelmelcher M, Becker SC, Donovan DM: Staphylococcal phage 2638A endolysin is lytic for *Staphylococcus aureus* and harbors an inter-lytic-domain secondary

translational start site. Appl. Microb. Biotechnol. 2013;97:3449–3456. doi:10.1007/s00253-012-4252-4

[39] Singh PK, Donovan DM, Kumar A: Intravitreal injection of the chimeric phage endolysin Ply187 protects mice from *Staphylococcus aureus* endophthalmitis. Antimicrob. Agents Chemother. 2013;58:4621–4629. doi:10.1128/AAC.00126-14

[40] Deresinski S: Bacteriophage therapy: exploiting smaller fleas. Clin. Infect. Dis. 2009;48:1096–1101. doi:10.1086/597405

[41] O'Flaherty S, Ross RP, Coffey A: Bacteriophage and their lysins for elimination of infectious bacteria. FEMS Microbiol. Rev. 2009;33:801–819. doi:10.1111/j.1574-6976.2009.00176.x

[42] Kutateladze M, Adamia R: Bacteriophages as potential new therapeutics to replace or supplement antibiotics. Trends Biotechnol. 2010;28:591–595. doi:10.1016/j.tibtech.2010.08.001

[43] Barrow PA, Soothill JS: Bacteriophage therapy and prophylaxis: rediscovery and renewed assessment of potential. Trends Microbiol. 1997;5:268–271. doi:10.1016/S0966-842X(97)01054-8

[44] Haq IU, Chaudhry WN, Akhtar MN, Andleeb S, Qadri I: Bacteriophages and their implications on future biotechnology: a review. Virol. J. 2012;9:9. doi:10.1186/1743-422X-9-9

[45] Berg-Miller ME, Yeoman CJ, Chia N, Tringe SG, Angly FE, Edwards RA, White BA: Phage–bacteria relationships and CRISPR elements revealed by a metagenomic survey of the rumen microbiome. Environ. Microbiol. 2012;14:207–227. doi:10.1111/j.1462-2920.2011.02593.x

[46] Shan T, Li L, Simmonds P, Wang C, Moeser A, Delwart E: The fecal virome of pigs on a high-density farm. J. Virol. 2011;85:11697–11708. doi:10.1128/JVI.05217-11

[47] De Paepe M, Leclerc M, Tinsley CR, Petit MA: Bacteriophages: an underestimated role in human and animal health? Front. Cell. Infect. Microbiol. 2014;4:39. doi:10.3389/fcimb.2014.00039

[48] Brüssow H, Canchaya C, Hardt WD: Phages and the evolution of bacterial pathogens: from genomic rearrangements to lysogenic conversion. Microbiol. Mol. Biol. Rev. 2004;68:560–602. doi:10.1128/MMBR.68.3.560-602.2004

[49] Tinsley CR, Bille E, Nassif X: Bacteriophages and pathogenicity: more than just providing a toxin? Microbes. Infect. 2006;8:1365–1371. doi:10.1016/j.micinf.2005.12.013

[50] Riley PA: Bacteriophages in autoimmune disease and other inflammatory conditions. Med. Hypotheses. 2004;62:493–498. doi:10.1016/j.mehy.2003.12.016

[51] Oviedo-Boyso J, Valdez-Alarcón JJ, Cajero-Juárez M, Ochoa-Zarzosa A, López-Meza JE, Bravo-Patino A, Baizabal-Aguirre VM: Innate immune response of bovine mam-

mary gland to pathogenic bacteria responsible for mastitis. J. Infect. 2007;54:399–409. doi:10.1016/j.jinf.2006.06.010

[52] Menzies PI, Ramanoon SZ: Mastitis of sheep and goats. Vet. Clin. North. Am. Food Pract. 2001;17:333–358.

[53] Sordillo LM, Streicher KL: Mammary gland immunity and mastitis susceptibility. J. Mammary Gland Biol. Neoplasia. 2002;7:135–146.

[54] Gill JJ, Pacan JC, Carson ME, Leslie KE, Griffiths MW, Sabour PM: Efficacy and pharmacokinetics of bacteriophage therapy in treatment of subclinical *Staphylococcus aureus* mastitis in lactating dairy cattle. Antimicrob. Agents Chemother. 2006;50:2912–2918. doi:10.1128/AAC.01630-05

[55] Pitkälä A, Haveri M, Pyörälä S, Myllys V, Honkanen-Buzalski T: Bovine mastitis in Finland 2001—Prevalence, distribution of bacteria, and antimicrobial resistance. J. Dairy Sci. 2004;87:2433–2441. doi:10.3168/jds.S0022-0302(04)73366-4

[56] Ángel-Andrés D, Solorio-Rivera JL, Baizabal-Aguirre VM, Bravo-Patiño A, Cajero-Juárez M, Sánchez-Gil LG, Chávez-Moctezuma MP, Valdez-Alarcón JJ: Genetic structure of *Staphylococcus aureus* population associated to bovine mastitis in backyard farms in Tarímbaro, Michoacán, by multilocus sequence typing (MLST). Biológicas. 2014;16:1–10.

[57] Carmona-Gasca CA, Talavera-Morales LE, García-Rodríguez FR, Orozco-Ortíz AP, Valdez-Alarcón JJ: NiS*spa*: open-code software for *Staphylococcus aureus* isolate typing by *spa*-typing for molecular epidemiology. Biológicas. 2012;14:30–36.

[58] Wieler LH, Antão EM, Semmler T: Research on zoonoses: central element of the "One Health" initiative. Berl. Munch. Tierarztl. Wochenschr. 2009;122:412–416.

[59] Graveland H, Duim B, vanDuijkeren E, Heederik D, Wagenaar JA: Livestock-associated methicillin-resistant *Staphylococcus aureus* in animals and humans. Intl. J. Med. Microbiol. 2011;8:630–634. doi:10.1016/j.ijmm.2011.09.004

[60] Valdez-Alarcón JJ, Bustos-Martínez JA, Baizabal-Aguirre VM, Bravo-Patiño A, Chávez-Moctezuma MP, Cajero-Juárez M: *Staphylococcus aureus* associated to bovine mastitis: a functional epidemiology approach. In: Velázquez-Ordoñez V, Castañeda-Vázquez H, Wolter W, Svarc GJ, Bedolla-Cedeño C, Guerra-Liera JE, editors. Production and quality of milk. Autonomous University of Sinaloa/Juan Pablo Editors; 2015. p 257–272. ISBN: 978-607-711-310-2. Ch16

[61] Franco A, Hasman H, Iurescia M, Lorenzetti R, Stegger M, Pantosti A, Feltrin F, Ianzano A, Porrero MC, Liapi M, Battista A: Molecular characterization of *spa* type t127 sequence type 1 methicillin-resistant *Staphylcoccus aureus* from pigs. J. Antimicrob. Chemother. 2011;66:1231–1235. doi:10.1093/jac/dkr115

[62] Sasaki T, Tsubakishita S, Tanaka Y, Ohtsuka M, Hongo I, Fukata T, Kabeya H, Maruyama S, Hiramatsu K: Population genetic structures of *Staphylococcus aureus*

isolates from cats and dogs in Japan. J. Clin. Microbiol. 2012;50:2152–2155. doi:10.1128/JCM.06739-11

[63] Gómez-Sanz E, Torres C, Benito D, Lozano C, Zarazaga M: Animal and human *Staphylococcus aureus* associated clonal lineages and high rate of *Staphylococcus pseudointermedius* novel lineages in Spanish kennel dogs: predominance of *S. aureus* ST398. Vet. Microbiol. 2013;166:580–589. doi:10.1016/j.vetmic.2013.07.014

[64] Hammad AM, Watanabe W, Tomoko F, Shimamoto T: Ocurrence and characteristics of methicillin-resistant and –susceptible *Staphylococcus aureus* and methicillin-resistant coagulase-negative *Staphylococcus* from Japanese retail ready-to-eat raw fish. Intl. J. Food. Microbiol. 2012;156:286–289. doi:10.1016/j.ijfoodmicro.2012.03.022

[65] Morris DO, Lautenbach E, Zaoutis T, Leckerman K, Edelstein PH, Rankin SC: Potential for pet animals to harbour methicillin-resistant *Staphylococcus aureus* when residing with human MRSA patients. Zoonoses Public Health. 2012;59:286–293. doi:10.1111/j.1863-2378.2011.01448.x.

[66] Gómez P, Lozano C, González-Barrio D, Zarazaga M, Ruiz-Fons F, Torres C: High prevalence of methicillin-resistant *Staphylococcus aureus* (MRSA) carrying the *mecC* gene in a semiextensive red deer (*Cervus elaphus hispanicus*) farm in Southern Spain. Vet. Microbiol. 2015;177:326–331. doi:10.1016/j.vetmic.2015.03.029

[67] Türkyilmaz S, Tekbiyik S, Oryasin E, Bozdogan B: Molecular epidemiology and antimicrobial resistance mechanisms of methicillin-resistant *Staphylococcus aureus* isolated from bovine milk. Zoonoses Public Health. 2010;57:197–203. doi:10.1111/j.1863-2378.2009.01257.x

[68] Capparelli R, Parlato M, Borriello G, Salvatore P, Iannelli D: Experimental phage therapy against *Staphylococcus aureus* in mice. Antimicrob. Agents Chemother. 2007;51:2765–2773. doi:10.1128/AAC.01513-06

[69] Kwiatek M, Parasion S, Mizak L, Gryko R, Bartoszcze M, Kocik J.Characterization of a bacteriophage, isolated from a cow with mastitis, that is lytic against *Staphylococcus aureus* strains. Arch. Virol. 2012;157:225–234. doi:10.1007/s00705-011-1160-3

[70] Mishra AK, Rawat M, Viswas KN, Abhishek, Kumar S, Reddy M: Expression and lytic efficacy assessment of the *Staphylococcus aureus* phage SA4 lysin gene. J. Vet. Sci. 2013;14:37–43.

[71] Hamza A, Perveen S, Abbas Z, Rehman, SU: The lytic SA phage demonstrate bactericidal activity against mastitis causing *Staphylococcus aureus*. Open Life Sci. 2016;11:39–45. doi:10.1515/biol-2016-0005

[72] Schmelcher M, Powell AM, Becker SC, Camp MJ, Donovan DM: Chimeric phage lysins act synergistically with lysostaphin to kill mastitis-causing *Staphylococcus aureus* in murine mammary glands. Appl. Environ. Microbiol. 2012;78:2297–2305. doi:10.1128/AEM.07050-11

[73] Becker SC, Roach DR, Chauhan VS, Shen Y, Foster-Frey J, Powell AM, Bauchan G, Lease RA, Mohammadi H, Harty WJ, Simmons C, Schmelcher M, Camp M, Dong S, Baker JR, Sheen TR, Doran KS, Pritchard DG, Almeida RA, Nelson DC, Marriott I, Lee JC, Donovan DM: Triple-acting lytic enzyme treatment of drug-resistant and intracellular *Staphylococcus aureus*. Sci. Rep. 2016;6:25063. doi:10.1038/srep25063

[74] Fan J, Zeng Z, Mai K, Yang Y, Feng J, Bai Y, Sun B, Xie Q, Tong Y, Ma J: Preliminary treatment of bovine mastitis caused by *Staphylococcus aureus*, with Trx-SA1, recombinant endolysin of *S. aureus* Bacteriophage IME-SA1. Vet. Microbiol. 2016;191:65–71. doi: 10.1016/j.vetmic.2016.06.001

[75] Matsuzaki S, Yasuda M, Nishikawa H, Kuroda M, Ujihara T, Shuin T, Shen Y, Jin Z, Fujimoto S, Nasimuzzaman MD, Wakiguchi H, Sugihara S, Sugiura T, Koda S, Muraoka A, Imai S: Experimental protection of mice against lethal *Staphylococcus aureus* infection by novel bacteriophage φMR11. J. Infect. Dis. 2003;187:613–624. doi:10.1086/374001

[76] Hsieh SE, Lo HH, Chen ST, Lee MC, Tseng YH: Wide host range and strong lytic activity of *Staphylococcus aureus* lytic phage Stau2. Appl. Environ. Microbiol. 2011;77:756–761. doi:10.1128/AEM.01848-10

[77] Takemura-Uchiyama I, Uchiyama J, Osanai M, Morimoto N, Asagiri T, Ujihara T, Daibata M, Sugiura T, Matsuzaki S: Experimental phage therapy against lethal lung-derived septicemia caused by *Staphylococcus aureus* in mice. Microbes Infect. 2014;16:512–517. doi:10.1016/j.micinf.2014.02.011

[78] Gu J, Zuo J, Lei L, Zhao H, Sun C, Feng X, Du C, Li X, Yang Y, Han W: LysGH15 reduces the inflammation caused by lethal methicillin-resistant *Staphylococcus aureus* infection in mice. Bioeng. Bugs. 2011;2:96–99. doi:10.4161/bbug.2.2.14883

[79] Gilmer DB, Schmitz JE, Euler CW, Fischetti VA: Novel bacteriophage lysin with broad lytic activity protects against mixed infection by *Streptococcus pyogenes* and methicillin-resistant *Staphylococcus aureus*. Antimicrob. Agents Chemother. 2013;57:2743–2750. doi: 10.1128/AAC.02526-12

[80] Schmelcher M, Shen Y, Nelson DC, Eugster MR, Eichenseher F, Hanke DC, Loessner MJ, Dong S, Pritchard DG, Lee JC, Becker SC, Foster-Frey J, Donovan DM: Evolutionary distinct bacteriophage endolysins featuring conserved peptidoglycan cleavage sites protect mice form MRSA infection. J. Antimicrob. Chemother. 2012;70:1453–1465. doi: 10.1093/jac/dku552

Bee Products and Essential Oils as Alternative Agents for Treatment of Infections Caused by *S. aureus*

Piotr Szweda and Barbara Kot

Additional information is available at the end of the chapter

http://dx.doi.org/10.5772/65978

Abstract

Bacteria of the genus *Staphylococcus* are important human and veterinary pathogens. A crucial characteristic for this group of bacteria is that they can easily acquire mechanisms of antibiotic resistance for a plethora of antibiotics currently in use for human and animal therapies. Therefore, there is a great need to find novel, non-antibiotic chemotherapeutics with marked antistaphylococcal activity. Promising but still underestimated group of potential antistaphylococcal chemotherapeutics constitute bee products: honey, pollen, royal jelly, fermented pollen and especially propolis. Another group of natural products that exhibit promising antibacterial activity is essential oils. Usefulness of bee products and essential oils in the treatment of infections caused by *S. aureus* has been confirmed by results of many investigations carried out by researches in different regions of the world. In this chapter, we have presented the review of publication in this area as well as perspectives and limitations of future applications of these two groups of natural products.

Keywords: *Staphylococcus aureus*, resistance, bee products, honey, propolis, pollen, fermented pollen, essential oils

1. Introduction

1.1. Staphylococci: important human and veterinary pathogens

Staphylococcus aureus is a species of bacteria commonly found in many ecological niches i.e., in soil, water and public places. These bacteria also colonize the skin and mucosal surfaces of humans and also several animal species. The most important site of *S. aureus* colonization is anterior nares of the nose, but these bacteria are also often isolated from the pharynx, perineum and axilla [1–3]. It was revealed in several independent studies that persistent colonization with *S. aureus* is observed in approximately 20% of human population, about 30% carry these microorganisms transiently and 50% are non-carriers [3–5]. *S. aureus* similarly as several other species of the genus *Staphylococcus* are classified as commensal Gram-positive bacteria. Like other commensal microorganisms, staphylococci have the ability to cause disease under certain conditions. It should be noted, however, that *S. aureus* is definitely one of the most dangerous commensal bacteria which colonize any part of the human body. Its high pathogenicity is based on production of a wide array of virulence factors such as protein A, coagulase, collagenase, hyaluronidase, hemolysins, lipases, multiple toxins, adhesive proteins and also proteins involved in biofilm formation. The ability to express these virulence factors has been confirmed not only for clinical isolates of *S. aureus,* but also for strains isolated from animal sources, e.g. bovine mastitis [6, 7] and food [8, 9]). All these isolates are potentially dangerous human pathogens. *S. aureus* has also developed several mechanisms that enable them escape from protective immune responses of infected humans or animals. Among them, protein A (SpA), staphylokinase and staphylococcal binder of immunoglobulin are the most important and the best characterized [10]. Another crucial characteristic for this group of bacteria is that they can easily acquire mechanisms of antibiotic resistance for a plethora of antibiotics currently in use for human and animal therapies. Especially important problem is rapidly growing number of isolation of strains resistant to methicillin, designated as *Methicillin Resistant Staphylococcus aureus* (MRSA). MRSA isolates are resistant to all β-lactam antibiotics. Moreover, some of them are classified as multidrug resistant (MDR)—not susceptible to the antibiotics that belong to various chemical groups. The MDR phenotype is usually associated with strains recovered from medical environment—*Healthcare Associated* MRSA (HA-MRSA). The drugs of choice for treatment of MRSA infections are glycopeptides, mainly vancomycin. Unfortunately, staphylococci developed also resistance to this group of antibiotics, and prevalence of isolation of vancomycin-intermediate *S. aureus* (VISA) and vancomycin-resistant *S. aureus* (VRSA) strains is constantly growing. The number of other antibiotics, which could be used for eradication of HA-MRSA infections, is very limited. In fact, only three antibiotics recently introduced to routine medicine procedures are effective in treatment of HA-MRSA infections: from the oxazolidinone group—linezolid; streptogramin group—quinupristin/dalfopristin, and from the glycylcycline group—tigecycline [11–13]. HA-MRSA are also usually susceptible to rifampicin, fusidic acid and co-trimoxazole; however, the last one does not give results good enough in treatment of acute infections, and in the case of two other drugs, resistance is easily acquired by treated bacteria [12–14]. The problem regarding therapy of staphylococci-related infections is additionally enhanced by the staphylococcal ability to form biofilm. Some of the previous studies proved that the biofilm-forming bacteria (including *S. aureus*) may be in fact 10–1000 times more resistant to antimicrobial agents than the same cells growing in a planktonic manner [15, 16].

Staphylococci also belong to the most important animal pathogens. Infections caused by these bacteria leads to huge economic losses in agriculture and food industry. Contamination of food products by these bacteria is a serious issue due to their completely lack of susceptibility to lysozyme and very low susceptibility to nisin, two important agents used as preservatives in food industry [17, 18].

http://dx.doi.org/10.5772/65978

1.2. Natural products as potential antimicrobial agents

All presented above aspects (the prevalence of the staphylococci in the environment, their high virulence potential and first of all rapid increase in antibiotic resistance) clearly indicate that there is an urgent need to develop new, effective inexpensive and not covered by current existing mechanisms of resistance antistaphylococcal agents. Interesting groups of antimicrobials, which meet all these expectations and could be used for treatment of *S. aureus* human and animal infections, but also for protection of food products against contamination by these bacteria, are bee products (honey, propolis, pollen, bee bread and royal jelly) and obtained from a broad spectrum of plant sources, essential oils (EO). Both these groups of products should be classified as well-known but rarely used antimicrobial agents. For centuries, herbs, bee products, venoms of some of animals (snakes and spiders) and other natural products were the only medicines that people knew and used. In some cases, the effects of the so-called folk/traditional medicine were surprisingly good, also in the case of therapy of infections. Unfortunately, from the beginning of "antibiotic era", they were nearly completely eliminated from clinical. One of the most spectacular examples of successful application of natural products for treatment infections is discovered by Prof. Tu Youyou, artemisinin. Artemisinin and its semi-synthetic derivatives exhibit the most rapid action against *Plasmodium falciparum* and are now standard treatment worldwide for *P. falciparum* malaria. For the discovery of artemisinin, Professor Youyou received the Prize of Nobel in 2015. The success of Prof. Youyou clearly confirms the high therapeutic potential of natural products, and it is also a good occasion for promotion of other natural products that exhibit antimicrobial activity but in fact are nearly completely forgotten, or rather eliminated, from clinical practice.

The main purpose of the preparation of this chapter is to demonstrate that bee and plant products are still interesting and promising group of antimicrobials. Their antibacterial activity is not only a matter of history from the pre-antibiotic era. Their large antimicrobial potential was confirmed with modern microbiological methods and analytical techniques and is well-documented in many scientific publications.

2. Antimicrobial activity of bee products

2.1. Honey

2.1.1. Basic information and mechanisms of antimicrobial activity

From ancient times, honey was a very valued component of the diet. This is mainly due to its characteristic sweet taste. The main uses of honey are as follows: as a spread on bread, a sweetener for tea, milk and coffee, for preparing desserts and cakes. Recently, it has been also proposed to use honey as a component of healthy, high energetic, beverages. The sweet taste of honey is apparent from its chemical composition. Sugars, mainly fructose and glucose, and minor amounts of oligosaccharides account for about 80% of its weight. The centuries of observations have also shown that regular consumption of honey is beneficial to the health of consumer. Therefore, this product was widely used as drug in traditional medicine, one of the most important except of herbs. The research carried out during several last decades finally confirmed that honey (and also other bee products) is a beneficial agent in treatment of a wide spectrum of human diseases, such as some cardiovascular and gastrointestinal tract disorders, infections within upper respiratory tract including cough, as well as in healing of infected wounds [19]. While folk medicine was based only on tradition and experience, the achievements of modern science and medicine have found confirmation of the therapeutic potential of honey, as the consequence of its chemical composition. It has been revealed that crucial for therapeutic properties of honey are its non-sugar components such as: enzymes, peptides, free amino acids, vitamins, organic acids, flavonoids, phenolic acids and other phytochemicals, minerals [20]. Especially interesting subject seems to be its antimicrobial activity. As it was mentioned above, this product was successfully used in treatment of skin and soft tissue infections and for elimination of pathogens infecting mucosal of respiratory tract. In all these situations, physical contact of honey (its components) with infected tissue (microorganisms infecting the tissue) is possible. The antimicrobial action of honey is based on several mechanisms: the acidity (low pH, usually in the range from 3.4 to 6.1), osmotic pressure of sugars present in honey and the presence of bacteriostatic and bactericidal substances such as H_2O_2, antioxidants, lysozyme, polyphenols, phenolic acids, flavonoids, methylglyoxal and bee peptides [21–24]. The crucial component that is responsible for the antimicrobial activity of majority of the honey types is hydrogen peroxide, which is formed as a side product of the oxidation of glucose by glucose oxidase—the enzyme which is introduced to the honey from the salivary glands of bees. Interestingly glucose oxidase is inactive in non-diluted honey [25]. The inhibition of growth of bacteria in honey, at the original concentration of this product, is mainly caused by high concentration of sugars (high osmotic pressure) coupled with high acidity [21, 22]. Therefore, it can be stored for long period of time (at least 2 years) without any additional treatment or supplementation with preservatives. When honey is diluted to certain extent, its antibacterial activity is shifted from osmotic- and pH-dependent to peroxide-dependent mechanism of action based on the generation of H_2O_2 [26]. This scenario takes place when honey is used for treatment of infections, e.g. infected wound. On the basis of current state of knowledge, it can be said that other mentioned above components of honey, mainly phytochemicals, seem to only support the antibacterial effect of generated H_2O_2. The dominant role of the enzyme in the antimicrobial activity of several polish unifloral honeys was recently confirmed in the researches that were carried out in our group. Preincubation of honeys solutions in 80°C for only 10 min resulted in complete loss of antibacterial activity of all tested honeys. The same effect was observed when suspensions of tested bacterial cells in solutions of honeys were supplemented with catalase. In both cases the observed, complete losses of activity of honeys were the consequence of lack of possibilities for H_2O_2 generation. Heat treatment resulted in denaturation of the enzyme, and in the presence of catalase, the generated hydrogen peroxide was immediately decomposed [27]. On the other hand, antimicrobial activity of honey also depends on the botanical source which was used by bees to collect the nectar. Buckwheat, thyme and cornflower honeys usually exhibit high antimicrobial activity, whilst produced in Poland in large amounts rape honey do not affect the growth of neither Gram-positive nor Gram-negative bacteria [27, 28]. Thus, the types of phytochemicals as well their concentration is important for final antimicrobial potential of honey. The issue of role of phytochemicals in the antimicrobial potential of honey is still not clear and is a subject of many interesting research. Hydrogen peroxide is also a known cytostatic agent; however, its concentration in honey is on a very low level which is safe for humans and animals. According to Lusby et al. [29], the concentration of H_2O_2 in honey is thousand times lower than in the common 3% antiseptic solution available in pharmacies; however, its constant production causes prolonged activity which can be considered as an advantage. The concentration of hydrogen peroxide on a non-toxic and stable level is probably regulated by antioxidants and pollen-derived catalase which destroy excess amounts of H_2O_2 [30].

2.1.2. *Honey as a potential antistaphylococcal agent*

The results of many investigations, carried out in different geographical regions of the world, revealed especially high efficiency of honey in treatment of infections caused by Gram-positive microorganisms, e.g. staphylococci. The mentioned above investigation carried out in our group revealed that honeys obtained from some species of plants, namely cornflower (*Centaurea cyanus* L.), buckwheat (*Fagopyrum esculentum* Moench) and thyme (*Thymus vulgaris* L.) were able to inhibit the growth of reference strain *S. aureus* PCM 2051 at the concentration of 3.12 or 6.25% (v/v). Satisfactory activity (minimal inhibitory concentration—MIC ≥6.25%) was observed for honeys obtained from linden tree (*Tilia* spp.), heather (*Calluna vulgaris* L.), savory (*Satureja hortensis* L.) and coriander (*Coriandrum sativum* L.). Other tested bacteria *Staphylococcus epidermidis* PCM 2118 and especially *Pseudomonas aeruginosa* ATCC 27853 and *Escherichia coli* K12 were less susceptible [27]. Quite similar results were obtained in the research carried out by the other polish group that investigates antimicrobial potential of bee products. Buckwheat, linden tree and heather honeys were found very active. Moreover, the authors also revealed high activity of goldenrod honey (*Solidago* spp.)—that was not observed in our research and honeydew honey—not investigated in our research. The latter is a specific kind of honey produced by bees, which collect the honeydew, the sweet secretions of aphids or other plant sap-sucking insects [28]. The observed in our studies activity against *S. aureus* was similar to the activity exerted by Slovenian chestnut, fir honeydew and forest honey (MIC = 2.5%, v/v) reported by Kuncic and coworkers [31] as well as Chilean Ulmo tree honey (MIC = 3.1%, v/v) reported by group of Sherlock [32]. The MIC values (against *S. aureus*) in the range of concentrations from 3.12 to 12.5% (v/v) were also observed by Anthimidou and Mossialos, who investigated a collection of 31 Greek and Cypriot honeys, they also revealed that Gram-negative bacteria *P. aeruginosa* revealed a bit higher resistance with MICs values in the range from 6.25 to 25% (v/v) [33]. Lallam and coworkers investigated antibacterial potential of 32 samples of honey (14 monofloral and 18 multifloral) collected from the Algerian Sahara Desert against four bacteria; *Bacillus subtilis, Clostridium perfringens, Escherichia coli* and *S. aureus*. The research confirmed high antimicrobial activity of honeys collected from this region; however, only disc diffusion method was used by the authors. All floral origins of honey showed antimicrobial activity against *S. aureus* but with rather similar reactions (9–10.5 mm), except with *P. persica*-based honey, whose activity was only 6 mm [34]. High anti-staphylococcal potential of honeys was also confirmed for clinical isolates of these bacteria, including MRSA strains. Effective inhibition of growth of MRSA isolates has been revealed in the case of mentioned above Chilean honey obtained from Ulmo tree [32], Malaysian melaleuca honey [35], some Thai honeys, especially from longan flower [36], Finland [37] and also many other geographical regions. Moreover, some authors revealed high activity of honey in eradication MRSA infections using *in vivo* models [38]. Considering honey as a therapeutic, antimicrobial agent the honey produced from the manuka bush (*Leptospermum scoparium*) indigenous to New Zealand and Australia deserves special attention. In contrast to majority of other nectar and honeydew honeys, the crucial factor responsible for the bactericidal activity of manuka honey is high concentration of 1, 2-dicarbonyl compound—methylglyoxal (MGO) in this product. High acidity and sugar concentration as well as hydrogen peroxide generation play in this case supporting roles [39, 40]. As a consequence, antimicrobial activity of this honey is not affected by heat treatment, catalase or proteolytic enzymes (hydrolysing glucose oxidase) [37, 40]. Several investigations *in vitro* confirmed high bactericidal, including antistaphylococcal activity of this honey [e.g. 27, 33]. Jenkins and Cooper revealed synergistic action between manuka honey and some antibiotics against MRSA and *P. aeruginosa* strains isolated from wounds [41]. Moreover, manuka honey has been successfully used for treatment of chronic wound infections caused by MRSA [42].

2.1.3. Perspectives and limitations of treatment staphylococcal infections with honey

From the point of view of possibilities of exploiting of the therapeutic potential of honey especially promising, and most realistic, seem to be application of this product as a component of wound dressing materials. Using honey to eliminate pathogens from infected wounds has a long tradition, which can be counted in hundred or even thousands of years. Currently, several companies specialize in production of dressings containing honey. However, many technical problems still have to be solved: (1) elimination of indigenous flora of honey, especially spores, without thermal or chemical treatment—results in deactivation of glucose oxidase and loss of antimicrobial activity, (2) inhibition of natural process of crystallization of honey, (3) large diversity of antimicrobial activity of honey—only the honeys with certified antibacterial activities should be used for medical applications. These problems could be partly solved by using manuka honey. However, research of many investigators revealed that many "classical honeys"—which activity is based mainly on the generation of hydrogen peroxide, exhibit even higher activity in comparison to manuka honey with a high content of methylglyoxal—550 mg/L [27, 28, 33]. Thus, they are also good candidates to be used in clinical practice as a component of dressings, ointments and creams, or direct application on the surface of infected skin, sores, diabetic foot, or mucous, e.g. in the oral cavity or genital tract.

2.2. Propolis

2.2.1. Basic information

The bees collect four products: nectar and honeydew for production of honey (source of carbohydrates), pollen (source of proteins) and propolis. Propolis is not a component of bees' diet; however, it is absolutely necessary for the proper development of bee colonies. This product, which is also called bee glue, is a natural resinous substance produced from plants' buds and exudates, modified by addition of bees' salivary secretions and wax. Similarly, as in the case of honey, propolis is a product of complex chemical composition. Some of its ingredients mainly polyphenols and flavonoids exhibit high antimicrobial activity. As a consequence, it is used as a hive disinfectant. Bees use propolis for elimination of pathogenic microorganisms from the walls of hive and cells of honeycomb, in which larvae develop (nest wells) and honey is stored [43]. It is important for prevention of development of such dangerous pathogens as *Paenibacillus larvae*, the bacteria responsible for American foulbrood, one of the most important diseases of bees [44]. Some authors also suggest that components of propolis, at least partially, affect the growth of *Varroa destructor*—most damaging parasite affecting honeybee colonies [45]. Because of its antimicrobial activity, propolis is also used by bees for mummification of larger pests such as mice invading the hive, which were killed as a result of the sting. Mummification prevents the decomposition of the body of the pest—development of pathogenic flora as well as generation of unpleasant odour. Due to its physical and chemical properties (high viscosity, low water solubility), propolis is also used as an important building material for sealing the hive or natural habitats of these insects (protection against wind or rain, stabilization of construction), which is also important for safety of bees' colonies.

2.2.2. Possibilities of application of propolis in therapy of bacterial infections

Similarly to honey, propolis was widely used in traditional medicine. The detailed history of using propolis in medicine and a discussion of perspectives of its future application have been recently presented by Silva-Carvalho and coworkers [46]. The chemical composition of this product depends on many factors: the geographical region where it was collected (the species of plants which were available for bees), season, weather conditions and many other. Some significant correlations were found primarily in the case of chemical composition of propolis and place of its isolation. On this basis, several different types of propolis have been proposed, e.g. poplar propolis, birch, green, red, "Pacific" and "Canarian". This classification still evaluates and new, different types of propolis are being recognized, e.g. Mediterranean or Portuguese [46]. At least, 13 different types of propolis have been identified in Brazil [47]. Because of differences in chemical composition the biological, including antibacterial, activity of different propolis samples may vary significantly. However, the carried out to date studies revealed that staphylococci and other Gram-positive bacteria are usually highly sensitive to this product collected in many different geographical locations. Propolis remains especially popular in the non-conventional medicine in Brazil as well as in other tropical countries, thus the product obtained in this region is well-characterized. The global market size of propolis was about 2300 tons in 2015 (it is established that it will increase to 2900 tons in 2021), and Brazil is the largest production and exporting country of this product (https://www.whatech.com/market-research/materials-chemicals/125806-world-propolis-industry-trends-share-size-2021-forecast-report). High anti-staphylococcal activity of propolis sourced from State of Paraná, in Brazil, was observed by Pamplona-Zomenhan and coworkers, the MIC_{50} and MIC_{90} for the 210 strains (162 MSSA and 48 MRSA) were both 1420 μg/mL [48]. The results of investigations performed by the groups of Fidoralisi [49] and Santana [50] indicated that propolis extracts might be effective against mastitis-causing *S. aureus*. The group of Fidoralisi observed reduction in *S. aureus* growth on average, 1.5 and 4 $\log_{10}$ times at concentration of propolis 200 and 500 μg/mL, respectively. At concentrations of 1000 μg/mL, all tested propolis samples reduced bacterial growth to zero [49]. The same effect was observed by Santana and coworkers, but only in BHI (Brain Hearth Infusion) medium. The authors noticed that in milk the bactericidal dose was at least 20-fold greater [50]. Interesting results were also published by Suleman and colleagues [51]. Some of 39 South Africa propolis samples exhibited much higher anti-staphylococcal activity in comparison to three tested samples of Brazilian propolis with MIC and MBC values of only 6 μg/mL [51]. Al-Waili and coworkers revealed that ethanolic extracts of propolis collected from Saudi Arabia (EEPS) and from Egypt (EEPE) effectively inhibited the growth of antibiotic resistant *E. coli, S. aureus* and *C. albicans* in single and polymicrobial cultures [52]. Strong antioxidant and antibacterial activity of propolis sourced from three different areas of Sonoran Desert in northwestern Mexico were confirmed in the research of Velazquez group [53]. The MIC against *S. aureus* of the most active sample (coming from Ures) was 100 μg/mL [53]. An antimicrobial effect of propolis harvested from honeybees in subtropical eastern Australia was investigated by Massaro and coworkers. The two tested propolis crude, ethanolic extracts showed bactericidal effects against *S. aureus* ATCC 25923 reference strain at the concentrations of 0.37–2.04 mg/mL [54]. Propolis produced in many Asian and European apiaries is also effective in elimination of staphylococci. The number of publications presenting results of biological properties of propolis sourced from this region has evidentially increased during the last decade. Because of limited size of this chapter only selected investigations can be described herein. Some promising results regarding antistaphylococcal activity of ethanolic extract of Polish propolis (EEPP) against methicillin-sensitive *S. aureus* (MSSA) and methicillin-resistant *S. aureus* (MRSA) clinical isolates have been recently revealed by Wojtyczka and coworkers. The investigated EEPP displayed varying effectiveness against twelve *S. aureus* strains, with MIC in the range from 0.39 to 0.78 mg/mL, determined by broth microdilution method, and minimal bactericidal concentration (MBC) of the EEPP ranged from 0.78 to 3.13 mg/mL. The disk diffusion assay revealed also that EEPP-enhanced antistaphylococcal activity of eight classical antimicrobial antibiotics, namely: cefoxitin, clindamycin, tetracycline, tobramycin, linezolid, trimethoprim+sulfamethoxazole, penicillin and erythromycin [55]. The same authors also investigated EEPP activity against biofilm-forming coagulase-negative *Staphylococcus* strains. The biofilm formation ability of all tested *S. epidermidis* strains was inhibited at EEPP concentrations ranging from 0.39 to 1.56 mg/mL [56].

Due to its health-promoting properties, propolis is widely used as a component of cosmetics, some food and beverages. In our opinion, these applications are not adequate to the biological properties of this product. The results of many presented above studies clearly indicate that propolis is a promising antimicrobial agent. Moreover, many other biological activities of this product have been described, including antioxidant, antiviral, antitumor, antifungal and immunomodulatory properties [46]. Large diversity of its chemical composition and consequently its biological activity eliminate propolis from clinical applications. Nevertheless, this limitation should be considered in therapy of serious, life-threatening infections (or other diseases), where the exact amount of biological active agent has to be used. *Staphylococci* are often responsible for many, not really serious, but bothersome infections. We would suggest more frequent use of propolis for treatment of this type of diseases. Currently, most of these infections are treated with antibiotics, which should be rather reserved for serious infections. It leads to the overuse of antibiotics and development of resistance phenomenon. In contrast

http://dx.doi.org/10.5772/65978

to antibiotics, propolis is a mixture of many antimicrobial components, thus development of resistance against this product (parallel resistance to all active constituents) is not possible. We also agree with other authors that it is necessary to continue research on correlation of biological activity and chemical composition of this product. Identification of crucial ingredients or their compositions could be used also for treatment serious, life-threatening diseases.

2.2.3. *Other bee products: pollen and royal jelly*

Investigation of antimicrobial activity of pollen, fermented pollen (bee bread) and royal jelly is not as popular as in the case of honey and propolis. However, several authors confirmed that these products also reveal some antimicrobial potential. Boukraa and coworkers revealed that MIC value for royal jelly against *S. aureus* is about 2% (v/v) [57], and the group of Gunaldi [58] when using this product obtained promising results in the preservation of implant-related infection in rats. The activity of royal jelly is probably a consequence of presence within composition of this product some peptides that inhibit growth of bacteria including members of the genus *Staphylococcus* [59].

Antimicrobial, antimutagenic, antioxidant and even anti-inflammatory activity of bee pollen (collected in Portugal and Spain) has been revealed in the research of Pascola and colleagues [60], and *S. aureus* was found as the most sensitive microorganism to the activity of these products. High antimicrobial activity of methanolic and ethanolic extracts of several monofloral Slovakian bee pollens was observed by the group of Fatracova-Sramkowa. *S. aureus* was found as most sensitive to the poppy pollen ethanolic extract. The most sensitive bacteria of rape bee pollen methanolic extract and sunflower ethanolic extract was *Salmonella enterica* [61].

2.2.4. *Honey and other bee products as a source of bacteriocinogenic bacteria with bactericidal antistaphylococcal activity*

However, bacteriocins are usually active against bacteria closely related to producing strain, but there are also many exceptions to this general rule, e.g. nisin, which is active against broad spectrum of Gram-positive bacteria [62]. Recently carried out research of the group of Prof. Worobo from Cornell University revealed that honey should be considered as a potential source of microorganisms producing promising antimicrobial compounds, especially bacteriocins [63]. The mentioned authors analyzed two Manuka honeys from New Zealand and six domestic honeys from US. The 2217 isolates out of 2398 strains (92.5%) exhibited activity at least against one of the tested microorganisms. Among all the bacterial indicator strains, *Listeria monocytogenes* had highest susceptibility (69%) to the various antimicrobial compounds produced by all active bacterial isolates (1655 out of 2398), whereas *S. aureus* showed the second highest susceptibility of all indicator microorganisms tested. Growth of this bacterium was inhibited by 66.9% of all the isolates (1605 out of 2398) [63]. The bacterial isolate TH13 was found as an efficient producer of peptide compound with high activity against *P. larvae* spp. *larvae*, which are etiological agents of American Foulbrood. It is a disease of honeybees that results in the annihilation of the honeybee colony [64]. Another isolate, identified as *Bacillus thuringiensis* SF316, was shown to efficiently produced thuricin H, a bacteriocin which strongly inhibits the growth of *Bacillus cereus* [65]. Thus, it was shown that antimicrobial activity of honey is attributed not only to hydrogen peroxide (formed by glucose oxidase originating from honeybees), antioxidant components or dimethyloglyoxal (in the case of Manuka honey), but also to microorganisms which are present in this product.

However, results of some successful investigations have been published the trials of isolation of bacteriocinogenic bacterial strains from honey and other than honeybee products were rarely carried out to date. In our opinion, especially, promising source of bacteria producing interesting bacteriocins could be fermented pollen—it is pollen which is collected by bees for the winter and early spring. The high antimicrobial activity of fermented pollen is the consequence of lactic acid bacteria (LAB) presence in this product, and the products of LAB metabolism—lactic acid, as well as bacteriocins.

3. Antibacterial activity of essential oils and plant extracts against *S. aureus*

Essential oils (EO) and their components are becoming increasingly popular as antimicrobial agents. They belong to the group of secondary metabolites that are enriched in compounds based on an isoprene structure and are called terpenes. They occur as di-, tri-, tetra-, hemi- and sesquiterpenes. The compounds that contain additional elements, usually oxygen, are termed terpenoids. Terpenes, terpenoids, as well as essential oils containing these substances exhibit antibacterial activity against broad spectrum of microorganisms including staphylococci [66].

One of the best characterized EO, which effectively inhibits growth of *S. aureus*, including MRSA isolates, is tea tree oil (TTO) derived mainly from the Australian native plant *Melaleuca alternifolia*. The antimicrobial activity of TTO is attributed mainly to its major component—terpinen-4-ol, and α-terpineol which is present in a lower concentration [67]. Several groups of researchers have evaluated the activity of TTO against MRSA. Carson et al. examined 64 MRSA isolates from Australia and the United Kingdom and showed that the MIC and MBC for the Australian isolates were 0.25 and 0.5% (v/v), respectively, while for the United Kingdom isolates were 0.312 and 0.625%, respectively [68]. The TTO has been also evaluated as an alternative decolonization agent for MRSA. The ointment and body wash containing 4 and 5% of TTO, respectively, were found as more effective in eradication of MRSA carriers than classical therapies with using 2% mupirocin nasal ointment and triclosan body wash [69]. Carson and coworkers revealed that TTO and its main components compromise the cytoplasmic membrane of *S. aureus* [70]. The most important consequences of the damages of lipid bilayer are as follows: leakage of important cytoplasmic components, inhibition of respiration (leakage of potassium ions), loss of sodium chloride tolerance and some changes in cell morphology [70, 71]. TTO, terpinen-4-ol and α-terpineol showed strong activity against biofilm formed by *S. aureus* on biomaterials. Partial destruction of 24-h-old biofilms was achieved in the concentration 4–8 times greater than MIC after 1 h, whereas 2–4 × MIC was adequate to obtain 90% reduction metabolic activity of biofilm after 4 h of treatment [72]. The research of some authors showed transient decreases in antibiotic susceptibility in several bacteria that had been exposed to TTO. It raises concerns that TTO may hinder the effectiveness of conventional antibiotics and influence the development of resistance [73–75].

http://dx.doi.org/10.5772/65978

However, recently Hammer et al. showed that the presence of TTO or terpinen-4-ol resulted in only minor changes in antibiotic susceptibility of *S. aureus* isolates that were serially sub-cultured with sub-inhibitory TTO or terpinen-4-ol [76].

Promising results were also obtained in the investigation of antimicrobial potential of lavender oil (LO). The oil obtained from *Lavandula angustifolia* demonstrated *in vitro* activity against MRSA at concentration of <1% [77]. Several chemically characterized lavender oils were assessed for their antibacterial activity using the disc diffusion method. All tested lavender oils inhibited growth of both MSSA and MRSA with inhibition zones ranged from 8 to 30 mm in diameter at oil doses ranging from 1 to 20 μL, respectively [78]. Some significant differences in the chemical composition and antibacterial activity of LO, which mainly depend on the origin of the lavender samples, were observed in research carried out by different authors. For example, it was demonstrated that oil from lavender of Bulgarian origin, which contains 51.1% linalool and 9.5% linalyl acetate as main components, was more effective against bacteria than oil originated from lavender sample of French origin containing 29.1% linalool and 43.2% linalyl acetate. It is believed, however, that lavender oil may be useful, first and foremost, as a prophylactic or topical application for surface infection [79].

Another interesting, from the point of view of its antimicrobial properties, is Thymus essential oils (TOs). Its main chemical components are α-thujone, α-pinene, camphene, β-pinene, p-cymene, α-terpinene, linalool, borneol, β-caryophyllene, thymol and carvacrol [80]. Different chemotypes of the essential oil from the genus *Thymus* were distinguished based on the presence of chemical components [81]. The antimicrobial properties of TOs are related to their high content of carvacrol and thymol, which were identified as the most efficient against bacteria [80]. Using an agar dilution method, the MIC values for MSSA and MRSA were reported for carvacrol (0.015–0.03%, v/v) followed by thymol (0.03–0.06%, v/v) [82]. The *Thymus* essential oils blended, in which the principal components were thymol, linalool, terpinen-4-ol and α-terpinene, exhibited significant inhibitory and bactericidal effects against strains of epidemic MRSA. The mean MIC and MBC values for the oil blend was 0.3 and 0.6%, respectively, whereas for the linalool chemotype thyme oil the MIC and MBC values were 0.4 and 0.8%, respectively. In the disc diffusion assay, the essential oils blended resulted in the formation of mean zone of inhibition size of 34.8 mm, while linalool chemotype produced a mean zone of 20.7 mm [83].

Recently research of many authors showed also the antimicrobial activity of geranium oil (GO) against MRSA [84, 85]. Among 67 components of geranium oil from *Pelargonium graveolens* Ait, citronellol, geraniol, nerol, citronellyl formate, isomenthone and linalool are the main constituents responsible for its biological activity. The research based on agar dilution method showed that the geranium oil had very strong activity against the clinical *S. aureus* strains, including MRSA strains, exhibiting MIC values of 0.25–2.50 μL/mL [85]. Moreover, Rosato et al. showed the occurrence of a synergism between geranium oil and norfloxacin against reference *S. aureus* strains [86].

The antibacterial activity of essential oils from oregano (*Origanum vulgare*) against multiresistant bacteria, including MRSA, was analysed by Costa et al. [87]. The MIC values were determined by the microdilution method. MRSA were inhibited by the essential oil at the concentration of 0.125%. Nostro et al. investigated activity of essential oils from oregano against biofilm-grown *S. aureus* and the effects of the oil on biofilm formation. The biofilm inhibitory concentrations (0.125–0.500%, v/v) and biofilm eradication concentrations (0.25–1.0%, v/v) were twofold or fourfold greater than the concentration inhibitory planktonic growth. Sub-inhibitory concentrations of the oils from oregano prevented biofilm formation by *S. aureus* strains [82].

Nigella sativa is a herbaceous plant cultivated in many countries in the world [88]. Crude extract and seed essential oil possess antibacterial activity against several bacteria [89]. The antibacterial effect may be due to the presence of the two important active compounds of *N. sativa*, thymoquinone and melanin [90]. The activity of *N. sativa* extract against clinical isolates MRSA was investigated by Hannan et al. [91]. They showed that all MRSA isolates were sensitive to *N. sativa* extract at a concentration of 4 mg/disc and MIC was in the range of 0.2–0.5 mg/mL. On the other hand, the multidrug resistant *S. aureus* strains isolated from nasal and milk samples of cows and buffalo were completely inhibited by *N. sativa* extract at concentration of 40 μg/mL on disc and MIC values were between 0.3 and 2.5 mg/mL [92].

Essential oil of cinnamon and cinnamaldehyde, which is main chemical constituent of this oil, also showed activity against MRSA. Essential oil from *Cinnamomum osmophloeum* (clone B) had an excellent inhibitory effect with the MIC of the essential oil and cinnamaldehyde against MRSA from human stand at 250 μg/mL [93]. The antimicrobial activity of cinnamon essential oil and *trans*-cinnamaldehyde against *Staphylococcus* spp. from clinical mastitis of cattle and goats was not dependent on the antibacterial susceptibility profile. However, the best antimicrobial activity was showed with *trans*-cinnamaldehyde and this compound could be used in the treatment of mastitis [94].

The number of research on antibacterial properties of extracts from medicinal plants against MRSA increased in recent years. These researches are conducted in different countries and show that extracts of plants are rich source of unique phytochemicals with activity against MRSA. Among recently investigated plant was *Schinus areira* L., which grows naturally in Argentina, Peru, Bolivia and Northern Chile. The essential oil from leaves and fruits of two specimens of *S. areira* differ in chemical profile. The limonene-rich oil isolated from the leaves and fruits had potent antibacterial effect on MRSA. When using 3.2 and 15 μL/mL (MICs value) of essential oil from leaves and fruits, respectively, the complete inhibition of MRSA growth was observed. Leaves and fruits oils showed bactericidal action after incubation for 24 h with 20 and 40 μL/mL, respectively. On the other hand, the α-phellandrene-rich fruit oil, having a lower content of limonene, was inactive against MRSA [95].

According to research of Endo and Dias Filho [96], MRSA is also sensitive to berberine (plant alkaloid) which is used in Chinese medicine. MICs values of berberine ranged from 62.5 to 250 μg/mL and MBC values were the same or twofold above the MIC. Highly potent anti-MRSA activity with MIC values in range of 25–50 mg/mL was detected among Libyan medicinal plants such as *Cistus salviifolius, Salvia officinalis, Pistacia atlantica, Arbutus pavarii* and *Myrtus communis* [97]. Significant anti-MRSA activity was documented in many studies on extracts of plants used in traditional medicine in Brazil. A mixture of hydrolyzable tannins from *Punica granatum* and the naphthoquinones α-lapachone I and α-xyloidone II from *Tabebuia avellanedae* showed antibacterial activity against all *S. aureus* strains tested, including MRSA isolates [98].

http://dx.doi.org/10.5772/65978

Turnera ulmifolia L. occurs in the north and northeast Brazilian regions and ethanol extract from this plant showed synergistic effect on gentamicin and kanamycin against MRSA strains. Coutinho et al. [99] found that the presence of ethanol extract of *T. ulmifolia* in growth medium at concentration of 32 μg/mL causes a significant reduction in the MIC for these antibiotics. The other studies conducted in India reported that ethanol, methanol and acetone extracts of *Moringa oleifera, Elettaria cardamomum* and *Tamarindus indica* seeds from India showed antibacterial activities against multidrug resistant MRSA isolates from wound infection [100].

4. Conclusions

Staphylococci belong to the most important pathogens for both humans and animals. The number of antibiotics effective in treatment of infections caused by these pathogenic bacteria is rapidly decreasing. Many centuries of observation and the use of bee products and essential oils in folk medicine as well as the results of advanced scientific research carried out during the last several decades clearly confirm high antimicrobial, including antistaphylococcal activity of these products. We have no doubt that they are an interesting and promising alternative to classical antibiotics and should be more seriously considered as therapeutic agents.

Acknowledgements

Preparing the chapter was supported by the Grant no 2015/18/E/NZ6/00700 from the "National Science Centre, Poland".

The authors are also grateful to Dr. Marta Schilemann for her help in preparing the English version of the manuscript of the chapter.

Author details

Piotr Szweda[1]* and Barbara Kot[2]

*Address all correspondence to: piotr.szweda@wp.pl

1 Department of Pharmaceutical Technology and Biochemistry, Faculty of Chemistry, Gdańsk University of Technology, Gdańsk, Poland

2 Department of Microbiology, Institute of Biology, Siedlce University of Natural Sciences and Humanities, Siedlce, Poland

References

[1] Williams REO. Healthy carriage of Staphylococcus aureus: its prevalence and importance. Bacteriol Rev. 1963;**27**:56–71.

[2] Kluytmans JA, Wertheim HF. Nasal carriage of Staphylococcus aureus and prevention of nosocomial infections. Infection. 2005;**33**:3–8.

[3] Wertheim HF, Melles DC, Vos MC, van Leeuwen W, van Belkum A, Verbrugh HA, Nouwen JL. The role of nasal carriage in Staphylococcus aureus infections. Lancet Infect Dis. 2005;**5(12)**:751–62.

[4] Eriksen NH, Espersen F, Rosdahl VT, Jensen K. Carriage of Staphylococcus aureus among 104 healthy persons during a 19-month period. Epidemiol Infect. 1995;**115**:51–60.

[5] Kluytmans J, van Belkum A, Verbrugh H. Nasal carriage of Staphylococcus aureus: epidemiology, underlying mechanisms, and associated risks. Clin Microbiol Rev. 1997;**10**:505–20.

[6] Gogoi-Tiwari J, Waryah CB, Eto KY, Tau M, Wells K, Costantino P, Tiwari HK, Isloor S, Hegde N, Mukkur T. Relative distribution of virulence-associated factors among Australian bovine Staphylococcus aureus isolates: potential relevance to development of an effective bovine mastitis vaccine. Virulence. 2015;**6(5)**:419–23. doi:10.1080/21505594.2015.1043508

[7] Kot B, Szweda P, Frankowska-Maciejewska A, Piechota M, Wolska K. Virulence gene profiles in Staphylococcus aureus isolated from cows with subclinical mastitis in eastern Poland. J Dairy Res. 2016;**83(2)**:228–35. doi:10.1017/S002202991600008X

[8] Kroning IS, Iglesias MA, Sehn CP, Valente Gandra TK, Mata MM, da Silva WP. Staphylococcus aureus isolated from handmade sweets: biofilm formation, enterotoxigenicity and antimicrobial resistance. Food Microbiol. 2016;**58**:105–11. doi:10.1016/j.fm.2016.04.001

[9] Baumgartner A, Niederhauser I, Johler S. Virulence and resistance gene profiles of Staphylococcus aureus strains isolated from ready-to-eat foods. J Food Prot. 2014;**77(7)**:1232–6. doi:10.4315/0362-028X.JFP-14-027

[10] Kim HK, Thammavon V, Schneewind O, Missiakas D. Recurrent infections and immune evasion strategies of Staphylococcus aureus. Curr Opin Microbiol. 2011;**15**:92–9. doi:10.1016/j.mib.2011.10.012

[11] Appelbaum PC, Jacobs MR. Recently approved and investigational antibiotics for treatment of severe infections caused by Gram-positive bacteria. Curr Opin Microbiol. 2005;**8**:510–7.

[12] Gilbert B, Robbins P, Livornese LL, Jr. Use of antibacterial agents in renal failure. Infect Dis Clin North Am. 2009;**23(4)**:899–924, viii. doi:10.1016/j.idc.2009.06.009

[13] Kurlenda J, Grinholc M. Alternative therapies in Staphylococcus aureus diseases. Acta Biochim Pol. 2012;**59(2)**:171–84.

[14] DeLeo FR, Otto M, Kreiswirth BN, Chambers HF. Community-associated meticillin-resistant Staphylococcus aureus. Lancet. 2010;**375**:1557–68. doi:10.1016/S0140-6736(09)61999-1

[15] Amorena B, Gracia E, Monzon M, Leiva J, Oteiza C, Perez JL, Alabart JL, Hernandez-Yago J. Antibiotic susceptibility assay for Staphylococcus aureus in biofilms developed in vitro. J Antimicrob Chemother. 1999;**44**:43–55.

[16] Melchior MB, Fink-Gremmels J, Gaastra W. Comparative assessment of the antimicrobial susceptibility of Staphylococcus aureus isolates from bovine mastitis in biofilm versus planktonic culture. J Vet Med B Infect Vet Public Health. 2006;**53**:326–32.

[17] Bera A, Herbert S, Jakob A, Vollmer W, Götz F. Why are pathogenic staphylococci so lysozyme resistant? The peptidoglycan O-acetyltransferase OatA is the major determinant for lysozyme resistance of Staphylococcus aureus. Mol Microbiol. 2005;**55(3)**:778–87.

[18] Szweda P, Schielmann M, Frankowska A, Kot B, Zalewska M. Antibiotic resistance in Staphylococcus aureus strains isolated from cows with mastitis in eastern Poland and analysis of susceptibility of resistant strains to alternative nonantibiotic agents: lysostaphin, nisin and polymyxin B. J Vet Med Sci. 2014;**76(3)**:355–62.

[19] Bogdanov S. The bee products: the wonders of the bee hexagon, bee product science, July 2011. www.bee-hexagon.net [accessed: 14 September 2014].

[20] Terrab A, Gonzalez AG, Diez MJ, Heredia FJ. Characterisation of Moroccan unifloral honeys using multivariate analysis. Eur Food Res Technol. 2003;**218**:88–95.

[21] Molan PC. The antibacterial activity of honey. 1. The nature of antibacterial activity. Bee World. 1992;**73**:5–28.

[22] Molan PC. The antibacterial activity of honey. 2. Variation in the potency of the antibacterial activity. Bee World. 1992;**73**:59–76.

[23] Mandal MD, Mandal S. Honey: its medicinal property and antibacterial activity. Asian Pac J Trop Biomed. 2011;**1**:154–60.

[24] Israili ZH. Antimicrobial properties of honey. Am J Ther. 2014;**21**:304–23.

[25] White JW, Subers MH, Schepartz AI. The identification of inhibine, the antibacterial factor in honey, as hydrogen peroxide and its origin in a honey glucose-oxidase system. Biochim Biophys Acta. 1963;**73**:57–70.

[26] Zainol MI, Yusoff KM, Yusof MYM. Antibacterial activity of selected Malysian honey. ComplementAltern Med. 2013;**13**:129. doi:10.1186/1472-6882-13-129

[27] Kuś PM, Szweda P, Jerković I, Tuberoso CI. Activity of Polish unifloral honeys against pathogenic bacteria and its correlation with colour, phenolic content, antioxidant capacity and other parameters. Lett Appl Microbiol. 2016;**62(3)**:269–76. doi:10.1111/lam.12541

[28] Kędzia B, Hołderna-Kędzia E, Dutkowiak A. The antibiotic activity of polish monofloral honeys. Postępy Fitoterapii. 2014;**2**:67–70 (in polish).

[29] Lusby PE, Coombes A, Wilkinson JM. Honey: a potent agent for wound healing. J Wound Ostomy Continence Nurs. 2002;**29**:295–300.

[30] Kędzia B, Honey: chemical composition and biological properties. Warszawa: PWR SA. 2008 (in Polish).

[31] Kuncic MK, Jaklic D, Lapanje A, Gunde-Cimerman N. Antibacterial and antimycotic activities of Slovenian honeys. Br J Biomed Sci. 2012;**69(4)**:154–8.

[32] Sherlock O, Dolan A, Athman R, Power A, Gethin G, Cowman S, Humphreys H. Comparison of the antimicrobial activity of Ulmo honey from Chile and Manuka honey against methicillin-resistant Staphylococcus aureus, Escherichia coli and Pseudomonas aeruginosa. BMC Complement Altern Med. 2010;**10**:47. doi:10.1186/1472-6882-10-47

[33] Anthimidou E, Mossialos D. Antibacterial activity of Greek and Cypriot honeys against Staphylococcus aureus and Pseudomonas aeruginosa in comparison to manuka honey. J Med Food. 2013;**16(1)**:42–7. doi:10.1089/jmf.2012.0042

[34] Laallam H, Boughediri L, Bissati S, Menasria T, Mouzaoui MS, Hadjadj S, Hammoudi R, Chenchouni H. Modeling the synergistic antibacterial effects of honey characteristics of different botanical origins from the Sahara Desert of Algeria. Front Microbiol. 2015;**6**:1239. doi:10.3389/fmicb.2015.01239

[35] Ng WJ, Lim MS. Anti-staphylococcal activity of melaleuca honey. Southeast Asian J Trop Med Public Health. 2015;**46(3)**:472–9.

[36] Jantakee K, Tragoolpua Y. Activities of different types of Thai honey on pathogenic bacteria causing skin diseases, tyrosinase enzyme and generating free radicals. Biol Res. 2015;**48**:4. doi:10.1186/0717-6287-48-4

[37] Huttunen S, Riihinen K, Kauhanen J, Tikkanen-Kaukanen C. Antimicrobial activity of different Finnish monofloral honeys against human pathogenic bacteria. APMIS. 2013;**121(9)**:827–34. doi:10.1111/apm.12039

[38] Medeiros Vde F, Azevedo ÍM, Rêgo AC, Egito ES, Araújo-Filho I, Medeiros AC. Antibacterial properties and healing effects of Melipona scutellaris honey in MRSA-infected wounds of rats. Acta Cir Bras. 2016;**31(5)**:327–32. doi:10.1590/S0102-865020160050000006

[39] Mavric E, Wittmann S, Barth G, Henle T. Identification and quantification of methylglyoxal as the dominant antibacterial constituent of Manuka (Leptospermum scoparium) honeys from New Zealand. Mol Nutr Food Res. 2008;**52(4)**:483–9. doi:10.1002/mnfr.200700282

[40] Kwakman PH, te Velde AA, de Boer L, Speijer D, Vandenbroucke-Grauls CM, Zaat SA. How honey kills bacteria. FASEB J. 2010 Jul;**24(7)**:2576–82. doi:10.1096/fj.09-150789

[41] Jenkins R, Cooper R. Improving antibiotic activity against wound pathogens with manuka honey in vitro. PLoS One. 2012;**7(9)**:e45600. doi:10.1371/journal.pone.0045600

[42] Visavadia BG, Honeysett J, Danford MH. Manuka honey dressing: an effective treatment for chronic wound infections. Br J Oral Maxillofac Surg. 2008;**46(1)**:55–6.

[43] Simone-Finstrom M, Spivak M. Propolis and bee health: the natural history and significance of resin use by honey bees. Apidologie, 2010, **41(3)**, 295–311.

[44] Bastos EM, Simone M, Jorge DM, Soares AE, Spivak M. In vitro study of the antimicrobial activity of Brazilian propolis against Paenibacillus larvae. J Invertebr Pathol. 2008;**97(3)**:273–81.

[45] Garedew A, Lamprecht I, Schmolz E, Schricker B. The varroacidal action of propolis: a laboratory assay. Apidologie. 2002;**33**:41–50.

[46] Silva-Carvalho R, Baltazar F, Almeida-Aguiar C. Propolis: a complex natural product with a plethora of biological activities that can be explored for drug development. Evid Based Complement Alternat Med. 2015;**2015**:206439. doi:10.1155/2015/206439

[47] Park YK, Alencar SM, Aguiar CL. Botanical origin and chemical composition of Brazilian propolis. J Agric Food Chem. 2002;**50**:2502–6.

[48] Pamplona-Zomenhan LC, Pamplona BC, da Silva CB, Marcucci MC, Mimica LM. Evaluation of the in vitro antimicrobial activity of an ethanol extract of Brazilian classified propolis on strains of Staphylococcus aureus. Braz J Microbiol. 2011;**42(4)**:1259–64. doi:10.1590/S1517-83822011000400002

[49] Fiordalisi SA, Honorato LA, Loiko MR, Avancini CA, Veleirinho MB, Machado Filho LC, Kuhnen S. The effects of Brazilian propolis on etiological agents of mastitis and the viability of bovine mammary gland explants. J Dairy Sci. 2016;**99(3)**:2308–18. doi:10.3168/jds.2015-9777

[50] Santana HF, Barbosa AA, Ferreira SO, Mantovani HC. Bactericidal activity of ethanolic extracts of propolis against Staphylococcus aureus isolated from mastitic cows. World J Microbiol Biotechnol. 2012;**28(2)**:485–91. doi:10.1007/s11274-011-0839-7

[51] Suleman T, van Vuuren S, Sandasi M, Viljoen AM. Antimicrobial activity and chemometric modelling of South African propolis. J Appl Microbiol. 2015;**119(4)**:981–90. doi:10.1111/jam.12906

[52] Al-Waili N, Al-Ghamdi A, Ansari MJ, Al-Attal Y, Salom K. Synergistic effects of honey and propolis toward drug multi-resistant Staphylococcus aureus, Escherichia coli and Candida albicans isolates in single and polymicrobial cultures. Int J Med Sci. 2012;**9(9)**:793–800. doi:10.7150/ijms.4722

[53] Velazquez C, Navarro M, Acosta A, Angulo A, Dominguez Z, Robles R, Robles-Zepeda R, Lugo E, Goycoolea FM, Velazquez EF, Astiazaran H, Hernandez J. Antibacterial and free-radical scavenging activities of Sonoran propolis. J Appl Microbiol. 2007;**103(5)**:1747–56.

[54] Massaro CF, Simpson JB, Powell D, Brooks P. Chemical composition and antimicrobial activity of honeybee (Apis mellifera ligustica) propolis from subtropical eastern Australia. Naturwissenschaften. 2015;**102(11–12)**:68. doi:10.1007/s00114-015-1318-z

[55] Wojtyczka RD, Dziedzic A, Idzik D, Kępa M, Kubina R, Kabała-Dzik A, Smoleń-Dzirba J, Stojko J, Sajewicz M, Wąsik TJ. Susceptibility of Staphylococcus aureus clinical isolates to propolis extract alone or in combination with antimicrobial drugs. Molecules. 2013;**18(8)**:9623–40. doi:10.3390/molecules18089623

[56] Wojtyczka RD, Kępa M, Idzik D, Kubina R, Kabała-Dzik A, Dziedzic A, Wąsik TJ. In vitro antimicrobial activity of ethanolic extract of Polish propolis against biofilm forming Staphylococcus epidermidis strains. Evid Based Complement Alternat Med. 2013;**2013**:590703. doi:10.1155/2013/590703

[57] Boukraa L, Niar A, Benbarek H, Benhanifia M. Additive action of royal jelly and honey against Staphylococcus aureus. J Med Food. 2008;**11(1)**:190–2.

[58] Gunaldi O, Daglioglu YK, Tugcu B, Kizilyildirim S, Postalci L, Ofluoglu E, Koksal F. Antibacterial effect of royal jelly for preservation of implant-related spinal infection in rat. Turk Neurosurg. 2014;**24(2)**:249–52.

[59] Romanelli A, Moggio L, Montella RC, Campiglia P, Iannaccone M, Capuano F, Pedone C, Capparelli R. Peptides from Royal Jelly: studies on the antimicrobial activity of jelleins, jelleins analogs and synergy with temporins. J Pept Sci. 2011;**17(5)**:348–52. doi:10.1002/psc.1316

[60] Pascoal A, Rodrigues S, Teixeira A, Feás X, Estevinho LM. Biological activities of commercial bee pollens: antimicrobial, antimutagenic, antioxidant and anti-inflammatory. Food Chem Toxicol. 2014;**63**:233–9. doi:10.1016/j.fct.2013.11.010

[61] Fatrcová-Šramková K, Nôžková J, Kačániová M, Máriássyová M, Rovná K, Stričík M. Antioxidant and antimicrobial properties of monofloral bee pollen. J Environ Sci Health B. 2013;**48(2)**:133–8. doi:10.1080/03601234.2013.727664

[62] Snyder AB, Worobo RW. Chemical and genetic characterization of bacteriocins: antimicrobial peptides for food safety. J Sci Food Agric. 2014;**94(1)**:28–44. doi:10.1002/jsfa.6293

[63] Lee H, Churey JJ, Worobo RW. Antimicrobial activity of bacterial isolates from different floral sources of honey. Int J Food Microbiol. 2008;**126(1–2)**:240–4. doi:10.1016/j.ijfoodmicro.2008.04.030

[64] Lee H, Churey JJ, Worobo RW. Isolation and characterization of a protective bacterial culture isolated from honey active against American Foulbrood disease. FEMS Microbiol Lett. 2009;**296(1)**:39–44. doi:10.1111/j.1574-6968.2009.01615.x

[65] Lee H, Churey JJ, Worobo RW. Biosynthesis and transcriptional analysis of thurincin H, a tandem repeated bacteriocin genetic locus, produced by Bacillus thuringiensis SF361. FEMS Microbiol Lett. 2009;**299(2)**:205–13. doi:10.1111/j.1574-6968.2009.01749.x

[66] Cowan MM. Plant products as antimicrobial agents. Clin Microbiol Rev. 1999;**12(4)**:564–82. Review.

[67] Carson CF, Hammer KA, Riley TV. Melaleuca alternifolia (Tea Tree) oil: a review of antimicrobial and other medicinal properties. Clin Microbiol Rev. 2006;**19(1)**:50–62.

[68] Carson CF, Cookson BD, Farrelly HD, Riley TV. Susceptibility of methicillin-resistant Staphylococcus aureus to the essential oil of Melaleuca alternifolia. J Antimicrob Chemother. 1995;**35(3)**:421–4.

[69] Caelli M, Porteous J, Carson CF, Heller R, Riley TV. Tea tree oil as an alternative topical decolonization agent for methicillin-resistant Staphylococcus aureus. J Hosp Infect. 2000;**46(3)**:236–7.

[70] Carson CF, Mee BJ, Riley TV. Mechanism of action of Melaleuca alternifolia (Tea Tree) oil on Staphylococcus aureus determined by time-kill, lysis, leakage, and salt tolerance assays and electron microscopy. Antimicrob Agents Chemother. 2002;**46**:1914–1920.

[71] Cox, SD, Mann CM, Markham JL. Bell HC, Gustafson JE, Warmington JR, Wyllie SG. The mode of antimicrobial action of the essential oil of Melaleuca alternifolia (tea tree oil). J Appl Microbiol. 2000;**88**:170–5.

[72] Budzyńska A, Wieckowska-Szakiel M, Sadowska B, Kalemba D, Różalska B. Antibiofilm activity of selected plant essential oils and their major components. Pol J Microbiol. 2011;**60(1)**:35–41.

[73] Nelson RR. Selection of resistance to the essential oil of Melaleuca alternifolia in Staphylococcus aureus. J Antimicrob Chemother. 2000;**45**:549–50.

[74] McMahon M, Blair I, Moore J, McDowell D. Habituation to sub-lethal concentrations of tea tree oil (Melaleuca alternifolia) is associated with reduced susceptibility to antibiotics in human pathogens. J Antimicrob Chemother. 2007;**59**:125–7.

[75] McMahon MA, Tunney MM, Moore JE, Blair IS, Gilpin DF, McDowell DA. Changes in antibiotic susceptibility in staphylococci habituated to sub-lethal concentrations of tea tree oil (Melaleuca alternifolia). Lett Appl Microbiol. 2008;**47**:263–8. doi:10.1111/j.1472-765X.2008.02420.x

[76] Hammer KA, Carson CF, Riley TV. Effects of Melaleuca alternifolia (Tea Tree) essential oil and the major monoterpene component terpinen-4-ol on the development of single- and multistep antibiotic resistance and antimicrobial susceptibility. Antimicrob Agents Chemother 2012;**56**:909–15. doi:10.1128/AAC.05741-11

[77] Nelson RR. In-vitro activities of five plant essential oils against methicillin-resistant Staphylococcus aureus and vancomycin-resistant Enterococcus faecium. J Antimicrob Chemother. 1997;**40(2)**:305–6.

[78] Roller S, Ernest N, Buckle J. The antimicrobial activity of high-necrodane and other lavender oils on methicillin-sensitive and -resistant Staphylococcus aureus (MSSA and MRSA). J Altern Complement Med. 2009;**15(3)**:275–9. doi:10.1089/acm.2008.0268

[79] Cavanagh HMA, Wilkinson JM. Biological activities of lavender essentials oil. Phytother Res. 2002;**16**:301–8.

[80] Cosentino S, Tuberoso CI, Pisano B, Satta M, Mascia V, Arzedi E, Palmas F. In-vitro antimicrobial activity and chemical composition of Sardinian Thymus essential oils. Lett Appl Microbiol. 1999;**29(2)**:130–5.

[81] Sáez F. Essential oil variability of Thymus hyemalis growing wild in Southeastern Spain. Biochem Syst Ecol. 1995;**23**:431–8.

[82] Nostro A, Sudano Roccaro A, Bisignano G, Marino A, Cannatelli MA, Pizzimenti FC, Cioni PL, Procopio F, Blanco AR. Effects of oregano, carvacrol and thymol on Staphylococcus aureus and Staphylococcus epidermidis biofilms. J Med Microbiol. 2007;**56**:519–23.

[83] Caplin JL, Allan I, Hanlon GW. Enhancing the in vitro activity of Thymus essential oils against Staphylococcus aureus by blending oils from specific cultivars. Int J Essent Oil Ther. 2009;**3**:35–9.

[84] Silva NCC, Fernandes Júnior A. Biological properties of medicinal plants: a review of their antimicrobial activity. J Venom Anim Toxins Incl Trop Dis. 2010;**16**:402–13.

[85] Bigos M, Wasiela M, Kalemba D, Sienkiewicz M. Antimicrobial activity of geranium oil against clinical strains of Staphylococcus aureus. Molecules. 2012;**17(9)**:10276–91. doi:10.3390/molecules170910276

[86] Rosato A, Vitali C, De Laurentis N, Armenise D, Antonietta Milillo M. Antibacterial effect of some essential oils administered alone or in combination with norfloxacin. Phytomedicine. 2007;**14(11)**:727–32.

[87] Costa AC, Santos BHC, Santos Filho L, Lima EO. Antibacterial activity of the essential oil of Origanum vulgare L. (Lamiaceae) against bacterial multiresistant strains isolated from nosocomial patients. Rev Bras Pharmacogn. 2009;**19(1B)**:236–41.

[88] Ahmad A, Husain A, Mujeeb M, Khan SA, Najmi AK, Siddique NA, Damanhouri ZA, Anwar F. A review on therapeutic potential of Nigella sativa: a miracle herb. Asian Pac J Trop Biomed. 2013;**3(5)**:337–52. doi:10.1016/S2221-1691(13)60075-1

[89] Salman MT, Khan RA, Shukla I. Antimicrobial activity of Nigella sativa Linn. seed oil against multi-drug resistant bacteria from clinical isolates. Nat Prod Radiance. 2008;**7(1)**:10–14.

[90] Bakathir HA, Abbas NA. Detection of the antibacterial effect of Nigella sativa ground seeds with water. Afr J Tradit Complement Altern Med. 2011;**8(2)**:159–64.

[91] Hannan A, Saleem S, Chaudhary S, Barkaat M, Arshad MU. Antibacterial activity of Nigella sativa against clinical isolates of methicillin resistant Staphylococcus aureus. J Ayub Med Coll Abbottabad. 2008;**20(3)**:72–4.

[92] Ejaz R, Ashfaq UA, Idrees S. Antimicrobial potential of Pakistani medicinal plants against multi-drug resistance Staphylococcus aureus. J Coast Life Med. 2014;**2(9)**:714–20.

[93] Chang ST, Chen PF, Chang SC. Antibacterial activity of leaf essential oils and their constituents from Cinnamomum osmophloeum. J Ethnopharmacol. 2001;**77(1)**:123–7.

[94] Dal Pozzo M, Loreto ÉS, Santurio DF, Alves SH, Rossatto L, de Vargas AC, Viegas J, da Costa MM. Antibacterial activity of essential oil of cinnamon and trans-cinnamaldehyde against Staphylococcus spp. isolated from clinical mastitis of cattle and goats. Acta Sci Veter. 2012;**40(4)**:1080.

[95] Celaya LS, Alabrudzińska MH, Molina AC, Viturro CI, Moreno S. The inhibition of methicillin-resistant Staphylococcus aureus by essential oils isolated from leaves and fruits of Schinus areira depending on their chemical compositions. Acta Biochim Pol. 2014;**61(1)**:41–6.

[96] Endo EH, Dias Filho BP. Antibacterial activity of berberine against methicillin-resistant Staphylococcus Aureus planktonic and biofilm cells. Austin J Trop Med Hyg 2015;**1(1)**:1005

[97] Abouzeed YM, Elfahem A, Zgheel F, Ahmed MO. Antibacterial in-vitro activities of selected medicinal plants against methicillin resistant Staphylococcus aureus from Libyan environment. J Environ Anal Toxicol. 2013;**3**:7. ISSN: 2161-0525.

[98] Machado TB, Pinto AV, Pinto MC, Leal IC, Silva MG, Amaral AC, Kuster RM, Netto-dosSantos KR. In vitro activity of Brazilian medicinal plants, naturally occurring naphthoquinones and their analogues, against methicillin-resistant Staphylococcus aureus. Int J Antimicrob Agents. 2003;**21(3)**:279–84.

[99] Coutinho HDM, Costa JGM, Lima EO, Falcão-Silva VS, Siqueira JP. Herbal therapy associated with antibiotic therapy: potentiation of the antibiotic activity against methicillin-resistant Staphylococcus aureus by Turnera ulmifolia L. BMC Complement Altern Med. 2009;**9**:13. doi:10.1186/1472-6882-9-13

[100] Karthy ES, Ranjitha P, Mohankumar A. Antimicrobial potential of plant seed extracts against multidrug resistant methicillin resistant Staphylococcus aureus (MDR-MRSA). Int J Biol. 2009;**1(1)**:34–40. doi:10.5539/ijb.v1n1p34